How Nature Works

School for Advanced Research
Advanced Seminar Series
Michael F. Brown
General Editor

Since 1970 the School for Advanced Research (formerly the School of American Research) and SAR Press have published over one hundred volumes in the Advanced Seminar Series. These volumes arise from seminars held on SAR's Santa Fe campus that bring together small groups of experts to explore a single issue. Participants assess recent innovations in theory and methods, appraise ongoing research, and share data relevant to problems of significance in anthropology and related disciplines. The resulting volumes reflect SAR's commitment to the development of new ideas and to scholarship of the highest caliber. The complete Advanced Seminar Series can be found at www.sarweb.org.

Also available in the School for Advanced Research Advanced Seminar Series:

The Psychology of Women under Patriarchy edited by Holly F. Mathews and Adriana M. Manago

Negotiating Structural Vulnerability in Cancer Control edited by Julie Armin, Nancy J. Burke, and Laura Eichelberger

Governing Gifts: Faith, Charity, and the Security State edited by Erica Caple James

Puebloan Societies: Homology and Heterogeneity in Time and Space edited by Peter M. Whiteley

New Geospatial Approaches to the Anthropological Sciences edited by Robert L. Anemone and Glenn C. Conroy

Seduced and Betrayed: Exposing the Contemporary Microfinance Phenomenon edited by Milford Bateman and Kate Maclean

Fat Planet: Obesity, Culture, and Symbolic Body Capital edited by Eileen P. Anderson-Fye and Alexandra Brewis

Costly and Cute: Helpless Infants and Human Evolution edited by Wenda R. Trevathan and Karen R. Rosenberg

Why Forage? Hunters and Gatherers in the Twenty-First Century edited by Brian F. Codding and Karen L. Kramer

Muslim Youth and the 9/11 Generation edited by Adeline Masquelier and Benjamin F. Soares

For additional titles in the School for Advanced Research Advanced Seminar Series, please visit unmpress.com.

How Nature Works

RETHINKING LABOR
ON A TROUBLED PLANET

Edited by Sarah Besky and Alex Blanchette

SCHOOL FOR ADVANCED RESEARCH PRESS • SANTA FE

UNIVERSITY OF NEW MEXICO PRESS • ALBUQUERQUE

© 2019 by the School for Advanced Research
All rights reserved. Published 2019
Printed in the United States of America

ISBN 978-0-8263-6085-4 (paper)
ISBN 978-0-8263-6086-1 (electronic)

Library of Congress Cataloging-in-Publication
data is on file with the Library of Congress.

Cover illustration: *Water Ways* created by
Meg Lemieur and Bri Barton, 2017
Designed by Felicia Cedillos
Composed in Minion Pro 10/14

The seminar from which this book resulted was
made possible by the generous support of the
Mill Foundation.

FOREWORD   vii
   Thomas G. Andrews

ACKNOWLEDGMENTS   ix

INTRODUCTION. The Fragility of Work   1
   Sarah Besky and Alex Blanchette

Part One. The Ends of Work

CHAPTER ONE. Exhaustion and Endurance in Sick Landscapes:
   Cheap Tea and the Work of Monoculture in the Dooars, India   23
   Sarah Besky

CHAPTER TWO. The Concentration of Killing: Soy, Labor,
   and the Long Green Revolution   41
   Kregg Hetherington

CHAPTER THREE. Making Monotony: Bedsores and
   Other Signs of an Overworked Hog   59
   Alex Blanchette

Part Two. Labor Struggles

CHAPTER FOUR. The Job of Finding Food Is a Joke:
   Orangutan Rehabilitation, Work, Subsistence,
   and Social Relations   79
   Juno Salazar Parreñas

CHAPTER FIVE. The Heat of Work: Dissipation, Solidarity,
   and Kidney Disease in Nicaragua   97
   Alex Nading

CHAPTER SIX. Metabolic Relations: Korean Red Ginseng and
   the Ecologies of Modern Life   115
   Eleana Kim

CHAPTER SEVEN. How Guinea Pigs Work:
Figurations and Gastro-Politics in Peru   131
María Elena García

CHAPTER EIGHT. Industrial Materials: Labor,
Landscapes, and the Industrial Honeybee   149
Jake Kosek

Part Three. Futures of Work

CHAPTER NINE. Cultural Analysis of Microbial Worlds   171
John Hartigan Jr.

CHAPTER TEN. Rhapsody in the Forest:
Wild Mushrooms and the Multispecies Multitude   191
Shiho Satsuka

CHAPTER ELEVEN. Kamadhenu's Last Stand:
On Animal Refusal to Work   211
Naisargi N. Dave

REFERENCES   225

CONTRIBUTORS   249

INDEX   251

The world I thought I knew has fallen to pieces, shattered beyond repair. Despair grabs me in its clutch whenever this grim realization presses down on me, driving me to distraction, cynicism, or, in my better moments, to closing ranks with kin and kindred spirits.

I can't help but feel myself glimmering with hope, though, after reading the collection of essays you hold in your hand and recalling the livening few days I spent with the editors and authors at the School for Advanced Research in Santa Fe, New Mexico, in 2016.

I am not an anthropologist, but I have long turned to anthropology for inspiration. As a historian who has worked for more than two decades to make sense of the fraught intersections between capitalists and workers, culture and nature, Indigenous peoples and settler-colonial societies, and, most recently, human and more-than-human beings, I continually come back to anthropology when I need to find questions I should ask, bodies of theory I should get to know, and methods worth adapting to my studies of the past.

The field of anthropology has always struck me as endlessly rich. And yet it is not where I usually turn when I need a pick-me-up.

Who would?

And this is what makes *How Nature Works: Rethinking Labor on a Troubled Planet* so exceptional: the book you are about to read offers an unexpected exercise in hope. These essays are subtle rather than shrill, purposeful but never programmatic, simultaneously heady and heartfelt. They sound no unitary call for reform or revolution.

Instead, they point in many directions. Urgently yet tentatively, with genuine humility and the freshest of thinking, they explore questions that everyone who cares about this fragile world of ours should be asking.

They ask us to trouble work as well as to refuse it—to tear down categories while probing the multifarious, deeply fraught ways in which human and other-than-human beings and entities collaborate to make both nature and culture. They implore us to reserve wonder and resist so many of the ontological and political snarls in which we have simultaneously limited ourselves and enmeshed ever-widening circles of "others," thus jeopardizing our collective existence.

None of us can say for certain whether these invitations to rethink work and nature can help us—the collective, superhuman *US* encompassed by Buddhist invocations to extend consideration to all sentient beings as well as Lakota teachings to consider the impact of our actions on everyone and everything to whom we are—find the resilience required to survive these dark days of exhaustion and extinction on a warming planet that is overworked and undervalued.[1]

At the very least, these essays light some of the paths we would do well to search out if we are to foster needful ways of knowing and being, working and enduring.

The world I thought I knew has been rent asunder. The book you hold in your hands, though, gives me hope that perhaps a world I want to know might still remain possible.

There is so much work yet to do—and even more that would be better undone.

—Thomas G. Andrews

Note

1. I do not use the Lakota phrase because of the critique of it offered by Francis White Bird, "Levels of Lakota Language: Part Two," *Lakota Country Times*, February 24, 2018, https://www.lakotacountrytimes.com/articles/lakota-country-times-344/. Accessed November 27, 2018. I am reminded here of Gerald Vizenor, *Manifest Manners: Narratives on Postindian Survivance* (Lincoln: University of Nebraska Press, 1999).

We would like to begin by acknowledging the School for Advanced Research in Santa Fe, New Mexico, and its Advanced Seminar series for making possible the conversations that led to this volume. There are few other places so intellectually nourishing. For their support during and after the seminar, we thank in particular Michael Brown, Paul Ryer, Leslie Shipman, Maria Spray, Nicole Taylor, and Carla Tozcano. In addition to contributing an exceptional chapter to this collection, Naisargi N. Dave played an important role in helping us organize the seminar. Thomas G. Andrews selflessly served as a discussant during the seminar. His comments and published writing contributed much to this volume. We also thank all of the participants for their patient generosity over the years of compiling this book. Two anonymous reviewers provided exceptional feedback that helped guide the manuscript through its final stages.

The dialogue that led to this volume was first initiated at the 2014 Society for Cultural Anthropology Biennial Meeting in Detroit on "The Ends of Work." Thanks to John Hartigan and Cori Hayden for organizing that truly remarkable and generative meeting. Early drafts of this volume were discussed in a fall 2016 graduate seminar on "Labor and Social Life" at Brown University. Multiple chapters were also workshopped in March 2017 at Brown, where the sharp critiques and guidance of all participants strengthened this book. We offer our heartfelt thanks especially to Jane Collins, Naisargi Dave, Amelia Moore, Jason Moore, Michelle Murphy, Alex Nading, and Bhrigupati Singh. At Harvard's Radcliffe Institute, a writing group on critical food studies helped us polish the book's introduction. For their incisive comments, we thank Julie Guthman, Lisa Haushofer, Susanne Friedberg, Allison-Marie Loconto, and Wythe Marschall.

We were fortunate to be able to organize three panels around this book's themes at the 116th Annual Meeting of the American Anthropological Association in 2017. This offered an opportunity to put chapters from this volume into dialogue with other scholars' pursuits at the intersection of labor and nature. In addition to various people whose work constitutes the chapters of this volume, we thank Danielle DiNovelli-Lang, Radhika Govindrajan, Karen Hébert, Cymene Howe, Marcel LaFlamme, Amelia Moore, Heather Paxson, Lesley Sharp, and Amy Zhang. Jessica Cattelino, Naisargi Dave, Shiho Satsuka, and

Paige West provided important commentaries on papers during those discussions. These panels led to a "Theorizing the Contemporary" forum at *Cultural Anthropology* online. We thank Dominic Boyer and Cymene Howe for shepherding that collection. Contributors to that forum helped shape the thinking that went into this volume. In addition to those named above, thanks to Maan Barua, Thomas Cousins, Jamie Lorimer, and Michelle Pentecost.

The Fragility of Work

SARAH BESKY AND ALEX BLANCHETTE

In Paraguay, expanding fields of fast-growing soy plants displace prior generations of crops and peoples. In Malaysia, captive orangutans are forced to learn
how to earn their meals, trained by dispossessed farmers whose own survival
is now tied to rehabilitating endangered primates. In the United States, honeybees' bodies and life cycles are reshaped to service sprawling almond monocultures. In India, farmers become suspicious that their own labor is morally
fraught and exploitative after a prized and prolific dairy cow suddenly refuses
to give them any more milk.

These scenes, drawn from this book, could be taken to exemplify a familiar
story of how capitalist production sweeps planetary ecologies like an autonomous and unstoppable force. They are certainly testaments to the unplanned
consequences and feedback loops that are ever more quickly passing across
human and nonhuman lifeways: the production of more feed for factory farms,
more species displaced at the edges of rainforests, more farmers stripped of
land, more peoples left with little to sell but their labor. This book, however,
argues that these cases also allow us to reflect on how relationships between
nature and labor are changing. They raise questions about the shifting participants, values, and rhythms underlying capitalist work; they lead us to ask who
(or what) should be included as a protagonist in the critical study of labor today.
Moreover, they also suggest an underexplored paradox: practices of labor in
service to capitalism are expanding—in the sense of being performed by more
diverse kinds of bodies and beings—at the exact moment that (human) work's
capacity to underpin and organize society seems to be waning.

How Nature Works: Rethinking Labor on a Troubled Planet is intended as an invitation for thought at a moment of conjoined economic and ecological precarity.
As a divergent collection of ethnographic writings, it asks how anthropologists

and allied scholars might best conceptualize the productive activity of diverse human and nonhuman beings in contemporary global capitalism. Some chapters in this volume address both the potentials and pitfalls of more-than-human labor studies, asking what it would mean to analyze industrial grains and animals as "workers." Others raise new questions about the shifting meaning of human work amid rising automation and technological inputs, as the actions of nonhumans—from tractor trailers, to breeding animals, to pesticides—increasingly condition the terms and rhythms of production. The chapters all debate the changing possibilities and limits of labor today, as both an analytical category and a historically privileged basis for fomenting political transformation.

Since the Enlightenment, the ability to do productive work has been framed as an exclusively human capacity—albeit one that has just as often functioned to justify hierarchies between peoples (see also Wynter 2003). Work has been a key means through which the aspirational European category of "Man" was formed and given flesh. John Locke (1980) argued that labor and the improvement of "nature" justified individuals' claims to land as private property. The displacement and murder of Native peoples deemed incapable of durable and transformative work was often underwritten by such ideologies (see Braun 2002). Karl Marx (1976) argued that while human beings plan their laboring endeavors, creatures such as honeybees or spiders do not. Human beings consciously make new worlds, leaving open the hopeful possibility for radical transformations in how they live together, while nonhuman others largely react to those changes (Marx 1976, 284). Both Marx's and Locke's understandings of work depend on fixed categorical divisions: nonhuman and human action, instinct and intention, objects and subjects, nature and culture.

Yet these philosophical divisions have come to feel tenuous—perhaps ironically right at their late industrial apotheosis when traces of human laboring activity suffuse every inch of the planet. The rapid instabilities wrought by climate change (Barnes and Dove 2015; Crate and Nuttall 2009), zoonotic disease (Nading 2015b; Porter 2013), species extinction (Van Dooren 2014; Rose 2013), oil and gas extraction (Appel 2012; Weszkalnys 2015), runaway toxic chemical and pesticide exposure (Graeter 2017; Guthman 2016; Shapiro 2015), nuclear radiation and fallout (Masco 2017), and landslides and environmental risk (Choi 2015; Kockelman 2016; Zeiderman 2016) all make evident that the strict conceptual division of the world into active working (human) subjects and passive worked-upon (nonhuman) objects is becoming more difficult to sustain. While many have considered how the end of pristine "nature" marks a new planetary epoch, the Anthropocene, what this means for the concept and practice of labor remains underexplored. In response, this book is motivated by the idea that ongoing planetary transformations require a critical reconsideration of labor as a key measure of human uniqueness, value, and social merit. If we accept that climate change and similar mutations are largely irreversible, then we might also take seriously the capacity of the nonhuman world to work upon us, against us, and perhaps with us.

These are more than just scholarly concerns. As we write, environmental instabilities are being actively linked in public discourse to a crisis in capitalist work. On the right wing, politicians tend to place environmental protection at absolute odds with the protection of wage labor. In the 2016 US general election, the slogan "Make America Great Again" indexed a call to unmake society in a host of different ways—from the overtly racist to the economically isolationist. But a core promise was a so-called return of manufacturing jobs from the Global South, and even the creation of new ones. Such economic growth, however, was to be fueled by the rollback of the environmental safeguards that were said to hold these jobs at bay. Across Europe and North America, people have witnessed a slow attrition of industrial jobs due to neoliberal outsourcing and the automation of manufacturing (Rifkin 1995; Kolbert 2016). Factories and mills have shut down (Dudley 1994; Collins 2003; Lamphere 1987), leaving those who once inhabited them constantly looking for work in an economy of zero-hour contracts, part-time shifts, and poorly paid food, hospitality, and service jobs (Walley 2013; Ehrenreich 2001; Jayaraman 2014; see also Graeber 2018). This change did not happen in the distant and forgotten past. Memories of past work loom large, while traces of industrial labor linger in landscapes and on bodies (Walley 2013; Muehlebach and Shoshan 2012). Yet the environmental consequences of fossil-fueled production aside, the fact is that many of those (often nostalgically) longed-for industrial jobs appear to be permanently gone. Embodied human labor has been supplanted, in part, by the actions of automated machines, microchips, herbicides, and genetically engineered life forms.

At first glance, the liberal response to twinned environmental and economic crises seems distinct. In the decade preceding the 2016 US elections, American center-left politicians vocally supported the creation of green jobs, the untapped potential of sustainably managed "ecosystem services," and even the promise of nuclear energy. The growth of the organic food movement and tax incentives for the purchase of solar panels, for instance, aim to confront environmental crisis through market mechanisms. These liberal responses may reject the conservative premise that the environment and the economy are at absolute odds, but they all have one thing in common: the unquestioned value of work. What is unsettling about even some of the most progressive policy approaches is that they rely on the assumption that the only way to solve planetary crisis is to put more and more human bodies to work. Residents of tropical islands, for example, must become eco-entrepreneurs, selling tourist experiences to make ends meet (A. Moore 2015). Others must put ideas of nature and ecological harmony to work in the sale of organic or fair-trade certified coffee, tea, and sugar in extremely competitive markets (Besky 2014; P. West 2012). Geo-engineering schemes that were unthinkable even a couple of decades ago, such as climate and solar management via injecting aerosols into the stratosphere, are now being seriously debated (McLaren 2015). Such projects promise a return to a pre-crisis planetary state through laboring innovation and will. Or, even more cynically, they use concern about ecological degradation to generate new futures of unending work (DiNovelli-Lang and Hébert 2018).

Across mainstream American political discourse, there is an enduring— and even intensifying—faith in the value of human labor at the exact moment when its capacity to organize society and mitigate environmental ills has never been more in question. In the face of robotics, rising unemployment, and "surplus" populations, more radical thinkers have started claiming that the end of (human) work is looming. They call for us to prepare for a postwork society that will require new ways of distributing time and resources, as well as new aspirations, values, and ideas of what it means to be human (Srnicek and Williams 2015; Harari 2017; McDermott Hughes 2017; Graeber 2018; see also Collins [2017] and Appel [2014] on projects to reimagine the economy). This book is therefore written at a moment of apparent global transition when the meaning and value of work is in flux—and with it, perhaps, long-standing interpretations of what it means to lead a life well lived. It is not at all clear, however, what a future world without work might entail (Atanasoski and Vora 2015; Federici 2018). Rather than forecasting fixed futures of work (or worklessness), then, we take utopian visions of abundance and stories of looming ecological disaster themselves as objects of analysis. The chapters of this volume describe fragile present-day worlds of work where no one certain end is inevitable.

Troubled Ecologies

As critical scholars of capitalism and the environment, we find ourselves at a crossroads. Along one path, we can analyze the contemporary situation through what one might term "disaster narratives." In such narratives, a capitalist world-system is preying on a vulnerable nonhuman nature, resulting in a warming planet, disease, and depleted resources. In this teleology, human work can only deplete a passive nature. A depleted nature, in turn, leads to more inequality and, ultimately, less work to go around.

It is equally tempting to go along a second path and embrace more hopeful eco-technical narratives, in which human ingenuity and laboring effort alone can bring a damaged ecology back into balance. In such narratives, not only is nature barred from producing value but human work itself becomes naturalized as the only means of saving nature. Here, work redeems nature, and nature redeems work.

A third and perhaps more radical path would be to question the very assumptions political and economic theory—left or right—have used to frame the relationship between nature and work. We might reject the notion that work is a uniquely human capacity and recognize that nonhuman beings (along with humans cast as not "fully human," such as enslaved peoples and domestic workers, or even nonbiological "things" like minerals and chemicals) have always produced value. What we think of as "nature" is that which the powerful refuse to acknowledge (and thus remunerate) as capable of real work (see also Federici 1975). Along this third path, the problem is not work per se but instead a human exceptionalism in how we define it. What remains, however, is an unshaken faith in work's capacity to redeem the world.

We have written this book because we do not find these three paths satisfying. They all, in different ways, reify work as unchanging and unchangeable. In the chapters that follow, we do not simply ascribe work to nonhuman actors (i.e., just saying that a flower naturally "works" when it photosynthesizes). Indeed, to say that all nonhumans naturally work risks the projection of capital's fixation on the value of human labor onto all of the planet's energies (see J. Moore 2015). This is well-trodden, fraught territory that is recognizable, for instance, within sociobiologists' efforts to project (and naturalize) human hierarchies and divisions of labor onto ant colonies (Gordon 2016). We are instead united by an ethnographic attention to moments when it is impossible to treat work as a natural feature or inborn capacity of any single life form. Work is not done by any *one* (see Andrews 2008, 125). This volume's aims are to both expand our thinking about why a renewed politics of work might matter for environmentalism and to experiment with ways to loosen work's grip on our thought. If acknowledging the agency of nonhumans is a key to survival on a damaged planet, we would argue that we also need to learn how to conceptualize those activities outside labor frames of creation, growth, and productivity.

To this end, this volume, in Donna Haraway's (2016) words, seeks to "stay with the trouble" stirred up by contemporary planetary fragility. For Haraway, the (European) idea of a stable, unified Nature is a foundational myth of technoscience and capital that must be unraveled if a politics suited to the contemporary moment is to arise. This notion of a timeless nature, in other words, has always been a fragile analytic category and a product of its times. In a world where nuclear disaster remains a possibility, industrial pollution is reshaping landscapes, and genetic engineering remakes some species even as others go extinct, what counts as "nature" is uncertain. Critical social scientists and humanists generally accept not only that nature is a complex analytical category but also that work has been a key part of these transformations (Cronon 1995; Smith 2008; White 1995a). Work, perhaps more than any other contemporary practice, has also been what brings people into contact with and allows them to cultivate specific knowledge of aspects of the nonhuman world. In the pages that follow, then, we aim to take that observation one step further through a kind of conceptual reversal: to put forth work as an analytical category that is modified and changing through its contacts with nonhuman nature. Like the "nature" shifting beneath our feet, work—and with it the idea of "productivity"—is a fragile category of practice tied to and changing within the worlds where it unfolds. Human labor is an ecological relationship; the efficacy of any act of labor is tied as much to the laboring environment as it is to human ingenuity or skill (see Andrews 2008).[1] We ask, then, not only how work changes nonhuman natures—but how the very idea of work, too, is changing.

With this in mind, the chapters in this volume each consider how humans and nonhumans occupy, make livelihoods from, and make sense of what we call "troubled ecologies." These are laboratories, factory farms, plantations, thinning rainforests, and militarized borders: spaces where the mutual fragility of nature and work become apparent. On the one hand, "troubled" here means an ecology in peril. We see spaces like the factory farm not as perfections of capital's capacity to exploit nature but as experimental (and remarkably unstable) projects on and with other beings. In spite of their long tenure and current global dominance of agricultural landscapes, things like soy, sugar, and tea plantation monocultures are far from stable systems of accumulation. They, too, require constant maintenance and care. These are sites that are exhausted and overworked but that also require ever more work to remain economically viable. On the other hand, these are spaces that "trouble" foundational analytic categories such as work and nature. The above sites are ecosystems that are, at the same time, labor regimes: landscapes that have been made to extract work, capitalisms that are themselves ecologies (J. Moore 2015). They make the boundaries between subjects and objects blurry, while forcing those who subsist in them to grapple with the inadequacy of ingrained capitalist concepts and binaries such as work or nature. In short, troubled ecologies call for new intersections of labor and environmental politics that are attentive to empirical specificities rather than well-worn analytical habits.

This book emerged from a five-day seminar convened at the School for Advanced Research (SAR) in Santa Fe, New Mexico, in 2016. The seminar aimed to explore what environmental scholarship might bring to the critical study of labor and what the feminist political theorist Kathi Weeks (2011) terms "postwork imaginaries." While there are copious critiques of *conditions* of work in journalism and scholarship, and many efforts to realize a world where work is justly remunerated and carried on with dignity, few question the institution of work itself, or the logic of a society where work disproportionately shapes everyday hierarchies and environments (but see Weber 2002; Gorz 1994; Ferguson 2015; Graeber 2018). Weeks (2011) is a notable exception, and the seminar was partly inspired by Weeks's call for attention to the ways that work—especially in the United States—has become depoliticized in the twenty-first century. Weeks diagnoses a curious kind of overriding ethic operating in the United States, describing it as a society dictated not only by capitalist labor value but by labor as a value. The United States is an economic and social locale where people must not only work to live but often appear willing to live for work (Weeks 2011, 2). As she puts it, "The social role of waged work has been so naturalized as to seem necessary and inevitable, something that might be tinkered with but never escaped" (7). Instead of denying the worth of actual workers, or the necessity of productive activity in general, she reminds us that "there are other ways to distribute or organize that activity" and "that is it also possible to be creative outside of work" (12). At a moment when work dominates Euro-American life, people spend much of their waking hours struggling to find and keep work, while the privileged few with stable employment are often overworked. Weeks thus asks us to consider how the hegemony of work circumscribes our political and intellectual imaginations.[2] Weeks's political project is to develop resources for imagining the end of work in two senses: decreasing the actual amount of time and energy that labor absorbs from human life and reducing the prominence of work in shared ideas about what it means to be fully human.

But for our fellow SAR seminar participants—anthropologists of the environment who pay attention to geographical, cultural, and ontological difference—Weeks's critique also raised questions: What do struggles within and against work ethics look like outside of the urban United States, including in various parts of the Global South? How are other-than-human beings made part of the material and imaginative logics of the work ethic, such as fast-growing industrial chickens whose genes and behaviors are being unendingly engineered for increased productivity? How might these beings literally naturalize work and its hegemony? To what extent is acknowledging other species' "work" central to imagining new forms of solidarity and changing (or sustaining) capitalism as we know it (see Coulter 2016; Battistoni 2017; Wadiwell 2018)? And how, at its most broad, might the politics and practices of realizing a post-work world be imagined outside of a strictly anthropocentric lens? In short, we started to ask what a postwork politics looks like when we do not assume that human beings are always and necessarily the solitary protagonists of labor. In thinking through the stakes of these questions—which revolve around critically expanding what we mean, following Weeks (2011), by the naturalization of work—we are inspired by three theoretical currents: anti-productivist Marxism's reading of capitalist labor value, feminist analyses of capitalism and what counts as work, and writing in science and technology studies (STS) and anthropology on interspecies relations.

Labor and Nature in Marxist Thought

In *Capital, Volume 1*, Marx describes labor as an open-ended and fundamental site of ongoing interaction between humanity and the natural world:

> Labour is, first of all, a process between man and nature, a process by which man, through his own actions, mediates, regulates and controls the metabolism between himself and nature. . . . Through this movement he acts upon external nature and changes it, and in this way, he simultaneously changes his own nature. (1976, 283)

As humans convert external nature into useful things, they "[develop] the potentialities slumbering within nature, and [subject] the play of its forces to [their] own sovereign power" (283). There is a complex relationality to these statements, as the nature of humans—including, presumably, the nature of their work—is constantly shifting alongside their transformations of the objects and ecologies that constitute their existence. For Marx, it is the capitalist wage relation that intensifies and makes the foundational separation between humans and nature seem inevitable, rather than historical (see Marx 1973, 488–90). It is only under capitalism, when labor itself is ossified into an abstract, buyable, measurable, quantifiable commodity that is the source of all value, that "we presuppose labour in a form in which it has an exclusively human characteristic" (Marx 1976, 283–84). Put differently, one might say that capital depends on the acceptance of the idea that labor is the exclusive property of the human. Capital is a cultural period that seizes human capacities to remake the world and makes them into the primary source of value that society pursues (see Graeber 2013). It is perhaps for this reason that distinct anti-productivist theorists such as Moishe Postone (1993) or Frank Wilderson III (2003) argue that we should not critique capitalism from the standpoint of labor—such as calling for liberation via unalienated work—but instead unpack how capital assigns primary value to labor.

In a sociohistorical system fixated on labor value, nature tends to be reduced to only two forms: as a raw resource destined to become a bearer of (labor) value or as materials shaped into industrial technologies that extract value from living people. Yet within Marx's writing there are also signs that the sociohistorical ideology of the capitalist system cannot completely contain, obfuscate, and blind us to the abiding relationalities of labor and nature. His concept of metabolism (*Stoffwechsel*) is also a means of interpreting the relational basis of ecological crisis:

> All progress in capitalist agriculture is a progress in the art, not of robbing the worker, but of robbing the soil; all progress in increasing the fertility of the soil for a given time is a progress toward ruining the more long-lasting sources of that fertility. The more a country proceeds from large-scale industry as the background of its development, as in the case of the United States, the more rapid is this process of destruction. (Marx 1976, 638)

Marx's ideas about the metabolic relationship of humans to nature, mediated through labor, have been foundational to the fields of political ecology, environmental history, and environmental anthropology. What is alarming about the processes that Marx describes is his prescient articulation of the finiteness of metabolic relationships: the inevitability that capital's expansion will meet an end. Subsequent sociological analyses of "metabolic rift" index the disruption of social and natural cycles under capitalism (Foster 1999; Schneider and McMichael 2010). Others suggest that ecological degradation is less a consequence of capitalist modes of production than a constitutive element of capitalism itself (J. Moore 2015; Besky, this volume). For geographer Jason W. Moore (2015), capitalism is not imposed on an ecology. It is itself an ecology, one that animates relationships between machines, grains, and chickens, along with shaping these discrete creatures in the flesh. Moore's perspective is built on a unique reading of capital's valorization process. While capital only recognizes human labor as the source of value, Moore argues, that very labor depends on the "unpaid energies" of nonhumans that are appropriated during the labor process. Environmental degradation is itself a matter of harnessing this "cheap nature" for accumulating capital. Nature is "cheap," Moore argues, because nonhumans and land (not entirely unlike domestic workers and slaves) have not been justly recognized and remunerated for their work in making capitalist labor possible.

Though the theoretical approaches to nature and labor in this volume vary, readers will notice that concepts from Marxist critiques of labor value and the systematic obfuscation of nonhuman beings remain central to our collective thinking. Some authors explicitly take up the question of metabolism (Kim, Nading, and Hartigan). Other contributors also use ethnographic findings to inquire into how processes of alienation, wages, and exchange relationships gain political traction in interspecies encounters (Parreñas, García, and Satsuka). Still others attend to how nonhuman beings are remade into elements of an overworked industrial infrastructure (Blanchette and Kosek). Just as Marx refers to inanimate machinery and the like as "dead labour that, vampire-like, only lives by sucking living labour, and lives the more, the more labour that it sucks" (1976, 342), several contributions explore the blurry divide between life and nonlife in troubled ecologies (Besky, Hetherington, Nading, and Dave).

Feminist Political Economy

This volume, however, is also an effort to think outside of capitalism's dominant categories, or, at the very least, to grasp those categories as objects of ongoing struggle on the ground. What the chapters all illustrate is not only that nature works (or is made to work) but also how this insight can enrich understandings of the diverse ways that human and nonhuman creative activities are understood. What is needed, we suggest, is a critical stance on what work (human or otherwise) is, or could be, in the first place. For this, we turn to feminist critiques of capitalism.

Feminist critiques have given rise to new and enriched ontologies of labor. Anthropologists were among the vanguard of feminist scholars pushing for an expanded concept of labor, both empirically and theoretically, focusing ethnographically on spaces outside the public sphere and the walls of the factory, while also tracking the increased role of women in the industrial workforce (di Leonardo 1987; Lamphere 1987; Ong 1987; Zavella 1987b). Central to this thinking was the idea that putative "non-work" spaces like those of the home and family do not, in Eric R. Wolf's terms, merely provide "compensatory functions, in restoring to persons a wider sense of identity beyond that defined by unitary demands of a job, be it cutting cane on a Puerto Rican plantation or tightening nuts on bolts in an assembly line" (2001, 172). Instead, feminist accounts initially framed unwaged domestic work as central to capitalism, insofar as this work reproduces labor-power. Indeed, the nuclear family can be said to have emerged alongside industrial capitalism (Collins and Gimenez 1990; Engels 2009; Harris 2000; H. Moore 1986; Tsing 2015). Later scholars extended these insights to highlight forms of work that were previously naturalized or ignored, including affective, emotional, and immaterial labor that has become more central than ever to the accumulation of capital (Hochschild 1983; Hardt and Negri 2000; Muehlebach 2011; Mankekar and Gupta 2016; Yanagisako 2012). There are some tensions in these moves to expand what counts as labor (see Dave, this volume). As Weeks (2011) argues, they come with the risk of cementing the value of labor by making it difficult to make political claims without invoking productive work as the basis of one's worth (see also Federici 1975).

Yet we would argue that feminist critiques have been equally responsible for destabilizing the sanctity of labor. By extending what work means, where it takes place, and where it is made—such that the ability of one person to stand on a factory floor is the product of a vast assemblage of contingent relations—feminist critiques foreground the shifting and unstable character of labor. On the one hand, these critiques do this through attention to the literal fragility of contemporary work. Feminist political movements have sought to defend bodies from precarious conditions of work, from homes surrounding polluted deindustrialized brownfields to the toxic atmospheres of office buildings (Brown 1995; Lamoreaux 2016; Murphy 2006; Jain 2006). In such accounts, neither laboring nor reproductive lives appear "natural" or even desirable but merely the contingent products of their environment. They could (and should) be otherwise.

On the other hand, many feminist critiques illustrate the fragility of capital's attempt to invest labor with inherent value. Feminist critiques do this by way of a refusal to propose a rescue of any kind of primordial nature from capitalism's labor processes. In Haraway's (1991) view, a feminist politics should not appeal to a fixed "unalienated labor," since all bodies are always already industrialized, formed historically, and worked upon: part animal, part machine. If there is not a pure nature—including a pure human nature—there can be no pure work either. Capitalism relies, in fact, on a host of "generative" activities that occur outside the space of wage relations and accumulative schemes yet is never fully innocent of them. Such activities include the construction of kin relations, the giving of gifts, the theft of property, the enslavement of people, the exploitation of animals, or militarization and the occupation of land (Kim, this volume; Bear et al. 2015; Tsing 2015; Gibson-Graham 2006).

A key contribution of the feminist critique of work, especially in anthropology, has therefore been to cut the seemingly monolithic power of capitalism down to size while simultaneously attending to the regional particularities and efficacies of labor struggle (Ong 1991). While Weeks's (2011) call for the development of utopian demands and imaginaries against work certainly shapes the discussions behind this volume, each contributor attends to the specific formations of capital, work, and nature that they encounter in a given milieu, along with the political possibilities that they present. This can lead to calls for the recognition of Indigenous and guinea pig labor in the Peruvian gastronomic boom analyzed by María Elena García, the reduction of labor in Alex Blanchette's reading of industrial hogs' overworked sensorium and hormones, or the wholesale refusal of work in Naisargi N. Dave's philosophical reimagining of animal life in India. What is crucial is that all chapters experiment with locally specific ways to politicize capital's fragile and desperate hold over work—either to make the sanctity of labor locally questionable or unpack the conditions that make it feel all too durable.

Science and Technology Studies, Nonhumans, and Nature

It bears repeating that throughout the book, nonhuman beings are not passive objects of the processes of doing work and composing nature. This orientation builds not only on feminist critiques of political economy but also on theory drawn from STS, a field that explores how the thing we know as "nature" is not a unified field but an unstable assemblage of agents. While seldom explicitly theorizing labor, this literature nonetheless illustrates how the work of (say) making scientific knowledge is always more than human: distributed among experts, technical auxiliaries, and policymakers as well as animals, plants, machines, and chemicals (Latour 1993, 2005; Stengers 2010; Jasanoff 2007; Law and Mol 2002). What counts as human knowledge at any time is contingent on the stability of such more-than-human arrangements.

In feminist STS, this stability is not only contingent on the conditions of the laboratory or the behaviors of human and nonhuman actors but also on the political and economic context in which they are brought together (Haraway 1991; Johnson 2015). Capitalism, colonialism, and racism are not incidental or external to the co-working capacities of humans and others. Much contemporary scientific labor, after all, is undertaken in an effort not just to produce basic knowledge but to produce capitalist value. Discussions in the anthropology of science and capital have been built around the question of how nonhumans' reproductive power is commodified. Terms such as "biocapital" (Sunder Rajan 2006), "promissory capital" (Thompson 2005; M. Fortun 2008), and "lively capital" (Haraway 2008; Sunder Rajan 2012) call attention less to "how nature works" under capitalism than to how the generative (reproductive) capacity of living beings (human and otherwise) is both determined by capitalist relations and captured by regimes of accumulation (Franklin and Lock 2003; Helmreich 2008).

Reading this literature, it is tempting to simply say that the reproductive or generative capacities of life are also work. In one sense, STS does develop insights about how capital has subsumed biological life into labor to such an extent that it is difficult to tell where one starts and another begins—not unlike the ways that everyday tasks, such as running a web search, are increasingly mined as data sources for corporate profits (Tadiar 2012; Johnson 2015). In another sense, this seems unsatisfying, because in this view, value can be seen to come from nature's fecundity and (re)productivity (see Paxson 2018). Such an approach not only tells us little about what it means to be one of the remaining (human) workers in an automated environment but also ignores feminist insights into how what we call "work" is always a process of struggle and ideology. Biotech is often thought of as labor reducing, in that it turns microbes, plants, or insects into "living machines" (Helmreich 2009; Hayden 2003; Carse 2014; Nading 2015a). Such projects, of course, necessitate a suite of human and nonhuman work. As Edmund Russell (2004) argues, biotechnological beings—such as an engineered salmon—are ambiguous. They trouble our inherited categories of industrialism. These beings are analyzable simultaneously as workers, as machines, or even as biological factories. The choice of which of these framings to emphasize is always a political one (see also Beldo 2017).

While STS has included nonhuman nature in discussions of capital, then, the question of how nature might be included in discussions of work has arguably been underexamined. The chapters in this volume explore the diverse ways that nonhuman working capacities have been explicitly considered and contested in different ethnographic locations. Here, we build on multispecies ethnography, which is rooted in STS, in four ways.

First, we deconstruct "work" in much the same way that this subfield has taken apart blinding abstractions such as "plant" or "animal" in order to become better attuned to the incredible diversity of beings (or practices) effaced by these labels (Kirksey and Helmreich 2010). Second, we ask how the stakes of key political demands, such as anti-work instigations, are complicated once the terrain of politics is stretched to include nonhuman others (de la Cadena 2010; Wolfe 2009). This allows us to question, for instance, why and to what ends anthropocentric assumptions make "the end of work" in popular discussions really just mean the end of *human* work. Third, we experiment with what it would mean to extend what were once considered purely human capacities like work across species lines, building on parallel efforts to rethink communication or culture in this manner (Kohn 2013; Hartigan 2014).

Fourth, and most crucially, we are indebted to multispecies ethnography's insistence that nonhumans are not just resources to consume, or symbols on which to build imaginative thoughts, but instead beings with whom we live (Haraway 2008; Lowe 2010; Tsing 2015). While many of the tools of multispecies ethnography have been developed in spaces of abundant possibility—in ecologies that may not be pristine but still radiate forms of more-than-human wildness—our focus on spaces of labor pushes us to engage living with the (semi)domesticated creatures of late industrialism such as cows, pigs, sugarcane, tea bushes, ginseng, guinea pigs, bacteria, honeybees, mushrooms, and orangutans. These are, in many ways, beings that have been—and still are being—shaped to work. This approach moves the conversation about nature and capital away from fixed categories of *who* and *what* and toward questions of "*where* and *when* — historical geographies marking the capture of living labor" (Barua 2017, 275).

Where and When, Here and Now

There is a final backstory to this book. Many of the participants in the Santa Fe seminar also attended the keynote discussion at the 2014 Society for Cultural Anthropology Biennial Meeting in Detroit, where the theme was "The Ends of Work." The meeting featured a keynote discussion between Kathi Weeks and anthropologist Sylvia Yanagisako. When the moderator asked both to describe their respective projects as succinctly as possible, their answers went something like this. Weeks said, "Political theorists ask: How do we get out of here?" By "here" she meant a society that is underpinned by conditions of exploitation, alienation, and inequality. The question for Weeks was how to imagine a better world: to chart a path out of capital's constitutive dynamics. Yanagisako's response to Weeks was, "Anthropologists ask: What do we mean by 'here'?" She emphasized methods for exploring lived relations between labor, capitalism, and inequality: an empirical and experimental project, rather than a general political theory that points directly toward one path. Put differently, she drew attention to the multiple "heres" of capitalism and alongside them the many potential lines of struggle and movement into other futures (Bear et al. 2015).

While the chapters in this volume might offer imaginative insights about what could come next, what unites many of them is that they strive to rethink work and nature through Yanagisako's question of what we mean by "here." Some chapters do engage questions about the endurance of work, the end of work, the refusal of work, and the exhaustion of work. Building on the long history of anthropological labor studies, each chapter is also firmly ethnographic. Ethnography has played a central role in pushing against abstract sociologies and economics that frame working people as homogenous "inputs" fueling capitalist accumulation, returning texture and meaning to scholarly depictions of laborers' lives (see Nash 1979; Salzinger 2003). The contributors of this volume, in turn, ask not only how metaphors and narratives about nature's capacity to work are historically situated but also how those very metaphors produce highly specific material realities. Furthermore, though it was not intentional, many of the chapters have come to focus on food. Producing, procuring, and preserving food are, after all, key acts of (re)productive work that humans and nonhumans alike seemingly cannot escape.

The chapters show that many workers across the globe, from animal rehabilitation workers to agricultural laborers, are asking themselves just these sorts of questions. Some chapters examine diverse ways of thinking about how contemporary nature is working. Other chapters explore hopeful alternatives: a life outside of capitalism's fixation on labor as value, or even outside work itself. In still other cases, the conditions of the present, in which work seems so precarious yet all encompassing, are hard to think outside of and overcome. A new politics of work not only expands categories but critically engages those categories. In line with Yanagisako's provocation to attend to the *here*, the way to do this is to approach the world without presuming to know what kinds of works or natures we might find in it. On a planet teeming with traces of labor, we emphasize that workers have political and philosophical reflections to make about the state of nature and the state of work. A more expansive labor politics, in other words, need not be only about the rights and remuneration of human workers alone.

The volume begins with three chapters on the uncertain "Ends of Work," charting the indeterminate place of the human body and living labor in industrialized spaces. They inquire into the state of labor in locales that purport to no longer require much work. Sarah Besky takes us into exhausted postcolonial tea plantations in India, charting the ways that tea plants' waning vitalities ripple through the bodies of immobile laborers. Here she examines the perplexing endurance of monocultural plantations in spite of their many precarities and injustices, developing an analytic vocabulary of work that departs from idioms of creative production to theorize the labor of maintaining troubled ecologies. Kregg Hetherington's chapter shifts us to a hotly debated site of global agribusiness: the soy fields of Latin America. Bulldozers have razed countless rainforest-dwelling species and pushed peasant farmers off their land. In these vast tracts of nearly human-less fields that seem to iconize forecasted future ecologies without work, Hetherington nonetheless theorizes pesticides and engineered seeds as dependent on an intensifying labor of killing that shows few signs of ending soon. From within the barns of American factory farms, Alex Blanchette examines the breeding of a fragile industrial pig whose existence is dependent on increasingly intimate, though invisible, forms of human work. The industrial pig is an animal whose very muscle fibers necessitate the exploitation of labor. Tracing emerging modes of relation to this being, Blanchette's chapter theorizes how highly automated spaces nonetheless create intense and even unprecedented forms of interspecies intimacy for the few workers who remain.

The middle five chapters offer creative reimaginings of "Labor Struggles." These include both anticipated forms of classed, raced, and gendered struggle over hierarchies in the workplace and an attention to how people strive for new understandings of the meaning and value of their own labor in troubled ecologies. Juno Salazar Parreñas's chapter brings readers to wrenching scenes of interspecies precarity in an orangutan sanctuary in Borneo, where dispossessed farmers debate what it means to train primates to work on their own to collect food. All the while these former peasants' own remuneration for rehabilitative work is shifting amid economic turmoil, where subsistence is no longer guaranteed for anyone. In this space of postindustrial ruin, dominant figures and frames of both labor and nature collapse as they flit across species. Alex Nading's contribution looks at what, at first glance, would appear to be a traditional site of labor struggle: Nicaraguan sugarcane plantation workers' efforts to organize amid an epidemic of deadly kidney disease potentially caused by brutal conditions of work within a warming climate. However, by retheorizing the ways that heat has figured into discussions of labor—as caloric energy or as offshoots of working bodies—Nading arches beyond a strict focus on workers' suffering and illuminates how they employ heat itself as a means of relearning both the nature of their work and shared conditions of solidarity.

Eleana Kim's contribution examines the rise of intensive agriculture along the highly occupied yet "pristine" Korean demilitarized zone, tracing how ginseng cultivation relates to contemporary imaginaries about urban work, energetics, and economic subjectivity. In so doing, Kim begins refashioning the concept of metabolic rift away from one of straightforward teleological despoliation toward on-the-ground struggles to manage the changing metabolisms of postindustrial plants and peoples. As a potent illustration of human labor's fragility, this chapter theorizes how much effort—agricultural, military, multispecies, and affective—underlies (and is effaced within) the seemingly simple act of growing ginseng. María Elena García's chapter reflects on the Peruvian metropolitan gastronomic boom and the Indigenous-reared *cuy* (guinea pig) bodies on which it has been built, opening up a struggle concerning how categories of "labor" and "nature" continue to take hold of distinct bodies in exclusionary and violent ways. Through a theorization of "figuration," this chapter engages the kinds of affective and representational labor fueling troubled ecologies, along with the ways that capital works to *erase* the work of others—both human and not human. Jake Kosek's chapter examines the conjoined exploitation of two kinds of migrant workers that underlie Californian fruit and nut fields: honeybees and beekeepers. The bee is a rare nonhuman being whose laboring capacities have never been doubted—it has long served as a model for grasping (and naturalizing) existing forms of human labor—and Kosek traces changes to this being's life cycle to provocatively retheorize the capitalist relation between work and commodity, or subject and object.

The volume concludes with three chapters that suggest other unseen "Futures of Work." These are not written in the mode of a prognostication but instead suggest conceptual openings for alternative politics of work that remain under-explored. They illustrate new ways of attuning to work as it unfolds in laboratories, forests, and farms today. John Hartigan's chapter brings us to a basic science laboratory in which the scientists examine, experiment with, and theorize the "work" of microbes. This chapter looks at science not in the context of capital and corporate power but as a means of opening up our understanding of culture and sociality. Arguing that we should not dismiss as mere metaphor the economic idioms in which scientists discuss bacterial motions, Hartigan uses these scientists' findings, such as how flagellar "motors" work with more power than any human-made engine, to question whether humans are really the iconic "workers" of the world. Among other interventions, his chapter opens potential modes of examining worldly work beyond the standard political economic frames of labor studies. Shiho Satsuka's exploration of matsutake mushrooms and their devoted collectors in Canada, in another direction, examines labor as an attunement to other species' rhythms that diverge strikingly from the familiar tempos of industrial capitalism. Developing a novel theorization of the "multispecies multitude" that underlies *all* work, Satsuka urges us to pay attention to nonhuman cooperative labors that exist in the world—along with the human workers who continue to attune to them in spite of industrial ruination. Naisargi N. Dave's playful philosophical exploration engages the open question of what an anti-productivist ecological politics could potentially entail. Theorizing human and nonhuman capacities to refuse work in scenes drawn from rural India, Dave's writing both pulls together latent conceptual threads from this volume and makes a world without work—for humans and nonhumans, in both practice and conceptual attachments—appear as an alluring, elusively vexing, and truly strange place.

Discussions of nature and work can often be polemical. They are about abrupt ends or utopian futures. On the left and the right, and from different theoretical and empirical corners of scholarship, one hears disjointed echoes of industrial disaster and eco-optimism—a descent into a world of too much work, and a future in which automation, ruination, and obsolescence mean that there is not enough work to go around. The accounts in the chapters that follow are different. They describe uncertain and often perilously fragile attempts to imagine new relationships between nature and work from within some of the most iconic spaces of late industrial despoliation on earth—troubled ecologies where there remains much work and vital happenings even as it is clear that things are not working out.

Notes

1. Along these lines, Haraway (2016) has also recently rethought "autopoetic" theories of work, which frame labor as a matter of individual creation, in favor of a "sympoetic" analysis that holds that any act of production is a matter of interdependent action (see also Gorz 1994).

2. Weeks's challenging of capitalist work ethics harkens back to classic anthropological discussions of peasant, plantation, and foraging life by Sidney W. Mintz (1960), June Nash (1979), Eric R. Wolf (2001), Marshall Sahlins (1972), and especially James C. Scott (1976), who framed efforts by marginalized people to work slower, to work badly, or simply to not work at all as a crucial form of resistance. While the hegemony of work remains sacrosanct in the United States and many other places, James Ferguson (2015, 21) notes that this does not necessarily mean that most people's actual lives are dictated by work. In 2012, the official unemployment rate—which measures how many are actively seeking jobs but cannot find them—in the United States was about 7 percent, yet only 58.6 percent of people living in the United States were actually working (including children and retirees). The remainder organized their lives in other ways.

Part One

The Ends of Work

CHAPTER ONE

Exhaustion and Endurance in Sick Landscapes
Cheap Tea and the Work of Monoculture in the Dooars, India

SARAH BESKY

Over the last five years, the Indian print media has reported on what appears to be a mounting epidemic of starvation and malnutrition on tea plantations in Northeast India. The coverage is accompanied by arresting visuals of emaciated bodies lying on the dirt floors of the small one- or two-room huts that house workers and their extended families. Ribs, eyes, and stomachs bulge. Tuberculosis treatment becomes sporadic, diabetes is unmanaged, and otherwise unremarkable intestinal infections go unmedicated. The plantations on which the workers in these stories live have closed. No work means no pay, and workers quickly run through their meager savings. Workers have to choose between food and medicine.

These stories describe the abandonment of plantation workers by capital. These stories rarely, however, point singular fingers of blame. The plantation just is. It is unquestioned; an entity beyond which it is hard to imagine economic and social life. In much of this public discourse, state- and national-level bureaucrats and politicians, plantation owners, labor unions, and even workers themselves are implicated in this crisis. Tea—as a product and a plant—is not. The tea plantation has been a fixture in Northeast India for nearly two centuries. It is a remnant of colonial control, but it produces a product that the vast majority of people living on the subcontinent and beyond consume not just every day but multiple times a day.

There is a distinct geography to this wave of abandonment. It is contained almost exclusively in an area called the Dooars, the narrow strip of land that hugs the India-Bhutan border (the northernmost part of the district of Jalpaiguri in West Bengal) and links India's Northeast with the rest of the country. The British annexed the region from the Kingdom of Bhutan in 1869. Almost immediately, they established tea plantations there. Today, the region is nearly completely blanketed by them.

These plantations produce a kind of tea known as CTC, an abbreviation for "cut-tear-curl." The international market is dominated by tea known as "orthodox": the leafy product that Euro-American consumers would readily recognize. The domestic market, on the other hand, is dominated by CTC. While other major tea-producing regions in India such as Assam and Kerala produce a combination of orthodox and CTC, in the Dooars, production is devoted almost exclusively to CTC. The state of West Bengal produces three hundred million kilograms of tea per year. Of this three hundred million, nine million is orthodox tea from the famed Darjeeling region, and the rest is CTC produced in the Dooars.

The term "CTC" comes from the process by which tea leaves are chopped up and rolled into tight little balls that pack the malty punch desired in the myriad regional variations on "chai." When people refer to tea as India's "national beverage," they are referring to kernels of CTC boiled together with milk and sugar (see Lutgendorf 2012). This beverage takes many forms and punctuates everyday life across an economically and socially diverse subcontinent. It is consumed in homes and at roadside stalls. It is carried out to fields in reused bottles and drunk out of dainty teacups in urban offices. In India, CTC is a staple along with rice, dal, oil, or salt.

When I came to the Dooars in late 2015, it appeared that the news stories were correct. A mass abandonment of tea plantations was under way. Over the previous two decades, dozens of tea plantations had closed. In 2014, sixteen plantations owned by just one company, Duncans Industries Limited, which was among the first to establish plantations in the region beginning in the 1870s, halted production. Companies like Duncans now claim that operating plantations has become too expensive. Tea bushes are old, overworked. The Nepali and Adivasi laborers who pluck leaves from those bushes are also aging. Companies claim that they can no longer afford to maintain either plants or laboring populations.

Importantly, however, plantations in the Dooars have not been *fully* abandoned. As the stories about starvation and sickness attest, many workers remain on the plantations. These workers have not simply resigned themselves to a life of deprivation and decay. Rather, as I found, many remain hopeful that companies like Duncans will reopen the plantations on which they—and oftentimes generations of family members before them—have spent their entire lives.

They do this with good reason. Legally speaking, many "closed" Dooars plantations have not been totally shut down. They are instead held in suspended animation. Under Indian law, tea plantation companies can pause production with the objective of getting their finances in order. In both governmental and informal discourse, such plantations are not considered "closed" but "sick." The sickness of plantations, as media accounts show, manifests itself in the sick and malnourished bodies of workers. And it is sickness, I argue, that allows for the plantation to persist unchallenged, unchanged.

A sick plantation is subtly marked. The gate to the factory is closed and locked. Walking past that gate, you are not met with the usual wafts of burning coal and freshly fired tea. Instead, the air is quiet and still. Beyond the factory, the fields are verdant. Tea bushes span out toward the horizon. People walk the paths that connect the fields with the factory and the villages of workers, but their head baskets are not filled with tea leaves as they would be on a healthy plantation but instead with laundry just washed in the river or grasses to feed livestock or to dry to make brooms to sell in town. Children play cricket in the paths between the houses. This might be the scene on an open plantation on a Sunday, the off-day for plantation workers, but on a sick plantation, this is the everyday. The sick plantation is occupied, but it is not worked.

From a critical perspective, plantation monocultures are always pathologi-cal, even when they are in full flower. Monocultures are extreme environments. Even in contemporary India, where plantation companies are no longer British owned, the monoculture retains what seems like an anachronistic structure. Workers reside year-round in plantation housing, and they depend not just on cash wages but on a combination of cash and in-kind benefits, including food and medical rations (Besky 2014). For individual workers, birth, mar-riage, illness, and death often occur in a single place: a company-owned village surrounded by acres of tea bushes.[1]

For Anna Tsing, monoculture is "landscape modification in which only one stand-alone asset matters; everything else becomes weeds or waste" (2015, 5). In Tsing's account, the makers of monoculture "dream" of alienation: "the ability to stand alone, as if the entanglements of living did not matter" (5). It is this dream of alienation that makes the system tend toward sickness. Tsing continues,

> When its singular asset can no longer be produced, a place can be aban-doned. The timber has been cut; the oil runs out; the plantation soil no

longer supports crops. The search for assets resumes elsewhere. Thus, simplification for alienation produces ruins, space of abandonment for asset production. Global landscapes today are strewn with this kind of ruin. (5–6)

In Tsing's account, the abandonment of monoculture appears almost inevitable. Eventually, pesticides stop working, carbon goes from the ground to the atmosphere, and weeds and waste take over. Some scholars have asked if the current era should be called not the Anthropocene but the "Plantationocene" (Haraway 2015). Monoculture is an overworked nature. To put it simply, plants, people, soil, and atmospheres in the world's monocultures all seem exhausted.

As a result of this exhaustion, small pockets of activists in the tea-growing regions in Northeast India have begun to advocate for the wholesale abandonment of the colonially rooted plantation system. I began looking at plantation closures in the Dooars thinking that I might find evidence of what Tsing has called "feral biologies": new multispecies relationships and alternative economies that tend to emerge in "capitalist ruins" (quoted in Haraway 2015, 159; Tsing 2015).

But abandonment is rarely a clean break; rather, it is a politically fraught and indeterminate process (Povinelli 2011; Fennell 2015). Nowhere is this more evident than in the Dooars. The "sick" plantation is a category somewhere between the "cheap nature" of a functioning monoculture and the feral state of an abandoned one (J. Moore 2016). A sick plantation is overworked, exhausted. It needs rest. What is sick can be healed, if only to eventually fall sick again. Cycles of plantation sickness and recovery are becoming something of a norm in the Indian tea industry, especially in those areas oriented to the production of CTC.

Plantation sickness reveals that attention to human-human and human-plant entanglements is key to understanding life not just in the ruins of monoculture but within monoculture itself. Despite dreams of alienation, monocultures "work"—when they work—by drawing capital, labor, and technologies together and keeping them there. The term "monoculture," then, is something of a misnomer. Work engenders a kind of "abnormal intimacy," to adopt a phrase from Patricia Zavella (1987a), between people and a wider landscape that might not "normally" meet outside of the context of labor. Agricultural work—even in industrialized monocultures—is not ever simply work on plants but with an array of human and nonhuman others, including plants but also houses, latrines, and coal-fired processing machinery (see also Blanchette

2015). Agricultural work is often maintenance work. This work can be exhausting while also cultivating forms of solidarity that endure.

As Elizabeth Povinelli has argued, endurance and exhaustion are conceptually intertwined. "The question of endurance—and its social antonym, refuses to consider the subject of being a secondary quality" (2011, 31–32). We cannot understand how monoculture fosters ecological and bodily exhaustion without also considering monoculture's equally remarkable capacity for endurance and coherence. Monoculture persists through tensions between endurance and exhaustion, intimacy and alienation. They persist through the work of maintenance. Such tensions are manifestations of the plantation's capacity to work, to exceed the sum of its parts. Povinelli continues on this: "Internal to the concept of endurance (and exhaustion) is the problem of substance: its strength, hardiness, callousness; its continuity through space; its ability to suffer and yet persist" (32).

It is this ability to "suffer yet persist" that concerns me here. I want to use Povinelli's emphasis on "the problem of substance" to examine the tea plantation's persistence in India. We can consider the plantations—particularly closed plantations—as kinds of "capitalist ruins": blasted landscapes unable to provide abundant sources of wealth for capital. But the plantation also highlights the question of home and belonging. Extracting capital from and making a home in monocultures are intertwined but separate processes. This story highlights the durability of the monoculture, and the recent epidemic of starvation highlights how difficult abandonment can be. Even in the apparent absence of work, monoculture endures and exhausts.

Monoculture and the Birth of Cheap Tea

As we often teach our introductory anthropology students, agriculture (or horticulture) marks human settlement. It is a sign that humans have stopped moving. They have grown roots. Intentionally cultivating plants is a means of making place and a reason to stay there. Plants conscript humans into relationships. Monoculture is what happens when "plants" become "plant." Think of the singularized English nouns "soy," "cotton," and "rubber."

To understand how nature works on twenty-first-century plantations, it is necessary to go back: to see how this particular form of human-plant intimacy gets put into motion and how it persists.

The first British-operated tea production in India in the early nineteenth century mimicked a Chinese model of "family garden farming," distinguished by small production plots where farmers grew green leaf tea and brought it to a centralized location for processing and packaging (McGowan 1860). British planters soon set out to "improve" on that that model by establishing vertically integrated plantations (McGowan 1860; see also Griffiths 1967). By the mid-1800s, planters had cleared forest, planted tea, conscripted laborers, and built on-plantation factories, all with an eye to making a faster, more efficient system for converting highly perishable green leaf tea to a fermented, dried, trans-portable (and drinkable) form (McGowan 1860; Baildon 1882, 30–34; Chamney 1930, 43–45).

Before they could build tea factories, however, the British had to create a monoculture. The anonymous author of *Tea Cultivation*, an 1865 instructional manual for would-be tea planters, outlines the precise steps that need to be taken to make a monoculture.[2] "Tea will grow better in virgin soil," the manual explains. "Village lands have long ago had all of 'the goodness' taken out of them." These lands are weedy, literally. The author observes that "the germs of all kinds of weeds, deposited by the wind, by birds, and in the cattle manure" sap the "strength or power of sustenance left in" formerly "cultivated" soil. Such soil we learn, requires five times as much labor "to keep." "The shade of the forest," on the other hand, "has checked vegetation, the wind has less power to deposit on the surface seeds of wild grasses, and perhaps, more than all, no manure (always so fruitful in the propagation of weeds) has been spread onto the land."

"It is then necessary," the author concludes after weighing the upfront and backend costs of planting on "village lands" against those of planting in "virgin soil," "to clear forest and jungle land."

"How is this best done?" he asks.

To complete the task of turning forest into monoculture, you must pay close attention to ecology. In eastern Bengal (the Dooars and Assam), as the author of *Tea Cultivation* explains, you must first cut the bamboo, grasses, and small trees, leaving them on the ground to dry for two to three months, rendering them into kindling. Next, bigger trees can be removed by "ringing," or carving a six-inch to two-foot circle of bark around the trunk. Sap circulates under tree bark, like blood under our skin. Ringing halts this process by choking the vascular flow of nutrients to the upper reaches of the tree. Ringing is a slow death.

Anticipating the would-be monoculturalist's concerns about the time and cost of this work, the *Tea Cultivation* manual warns about the dangers of the alternative: unchecked biological diversity. White ants make homes in large trees—even after they have been felled. The author opines, "Were I to commence a garden in Bengal, I would, in spite of the expense, cut down, cut up, and burn the big trees with all the others, and thus, I believe, much decrease the possible chance of damage from these insects." He adds that in the hills, the planter-to-be can save time by cutting these large trees up and "flinging them down" into the ravines.

Once bamboo and grasses have dried up, and trees have died, it is time to burn. According to the manual's author,

> It is a grand sight to see—the fire leaps along, urged by a strong wind, which is generally waited for, and the quantity of combustible material is so great that the moderate sized trees, which have been felled and which would not burn themselves, are completely consumed. A curious accompaniment to these fires is the sound emitted by the burning bamboos. It resembles incessant discharges of musketry. As I lay in bed one night, in the neighborhood of a blazing jungle, I might easily have fancied myself in action; in fact I did fancy so; for as sleep stole over me, the volley upon volley transported me to scenes far different from the evergreen tea gardens around.

The military overtones—flashbacks, of course, but also flash-forwards to the voracious flames of forests from California to Indonesia—are apt. Creating a monoculture is about eradication (of undesirable flora, fauna, and people), even if it is made in the pursuit of preserving some forms of (human) life. Eradication, however, is often incomplete. The work of monoculture is maintenance—working over and over again *toward* homogeneity but never quite achieving it.

When the Dooars was annexed from the Kingdom of Bhutan, British colonial officials deemed it a "non-regulated area," covered in forest populated by swidden cultivators. Before tea could be brought under cultivation, the Mech and Bodo people who lived on the areas deemed most desirable for tea had to be evicted. Within twenty years of the first tea planting in the Dooars, the vast majority of Mech and Bodo people living there were relegated to a 30.7 square-mile reservation near the town of Jalpaiguri (Debnath 2010, 138).

To staff Dooars plantations, planters conscripted Nepali and Adivasi people from outside the region. Deemed more suitable for plantation labor, these workers were housed in labor lines, simple wooden structures that later grew into plantation villages. Across the Dooars, the labor lines were gradually rebuilt as simple one- and two-room houses. Entire families lived and worked in these villages, comprising field labor and those working the machinery in plantation factories. Like the factories, domestic structures remained the property of plantation companies; indeed, they were as integral as machinery and fertilizers to the vertically integrated production infrastructure (Besky 2017).

Champions of imperial expansion hailed monoculture as a triumph of science and technology over "wild" landscapes and people. Vertical integration, in the form of the factory–field–domestic space combination of the plantation, was the sociotechnical apex of this colonial vision of efficiency in production. In this respect, tea was prototypical. Indian-grown tea, along with other colonial monocultures like sugar, coffee, and tobacco, formed an arsenal of "proletarian hunger-killers": the cheap energy that fueled the early carbon-based economy of British and American mills (Mintz 1979).

Low lying, with little shade and nearly no frost or snow, the Dooars offered tea planters a long growing season. This geographical location, however, affects the quality of finished tea. While plantations in nearby Darjeeling and Assam produced a delicate, aromatic, floral product prized by connoisseurs, Dooars plantations produced mass quantities of malty, tannic tea, marginal in taste and quality. Before Indian independence in 1947, much of this less-desirable Dooars tea would have been blended with the more floral teas from other regions to produce the tastes that European consumers desired. (Tea consumption within India itself was quite limited, contained to upper-class urban households.) After independence, Indian owners who gradually took over Dooars tea plantations from the 1950s to the early 1970s found themselves suddenly left with viable monocultures but without a market for their product.

In the 1960s, a new state agency called the Tea Board of India, the government regulator of the industry, began aggressively promoting domestic consumption. CTC processing machinery was key to this effort. Since it was designed to process tea quickly, without preserving the long cylindrical leaves prized by European drinkers, CTC machinery was perfect for high-output, low-quality monocultures like that of the Dooars. Installed in plantation factories across the Dooars, CTC machinery enabled what industry insiders describe as "continuous manufacture," reducing the number of machines that tea needed to pass through and thereby doubling the amount of finished tea a plantation could produce (see Griffiths 1967, 495–96). Since CTC is more pungent than leafy orthodox, what tea manufacturers call its "cuppage" is greater. In short, you can use less CTC to get the same kick of flavor, color, and caffeine. By the 1970s, Dooars tea was almost all processed in this way and almost wholly sold on the domestic Indian market. The little balls of CTC could be boiled with milk, water, sugar, and spices in a single pot at home, in roadside stalls, and even on railway platforms.

Monoculture and Maintenance

Monocultures are place based and climatically sensitive. The work of making tea plants (*Camellia sinensis* or *Camellia sinensis* var. *assamica*) grow in the Dooars has always been distinct from the work of making them grow in Darjeeling or Assam or Kerala. Still, as the author of *Tea Cultivation* explains, tea is less picky than most cash crops: "The tea plant is a very hardy one, and will *grow* in various climates and in almost all soils" (emphasis in original). The monoculture remains viable in part because of the endurance of the tea plant. It is tenacious, persistent, and forgiving of neglect or less-than-ideal placement.

Even if the plant is a hardy one, getting an "asset" to "stand alone" turns out to require lots of work (Tsing 2015). In the Dooars, keeping the Nepali and Adivasi laborers who do that work reasonably healthy and available to work requires an elaborate infrastructure: a company-run system for housing, clothing, feeding, and providing medical care. In order for plants to "stand alone," that infrastructure must itself be maintained. The plantation system was developed and refined to do that maintenance: to keep the ants, grasses, and trees at bay, and to keep houses, food stores, crèches, and infirmaries standing. Depending on the monoculture, maintenance work can include harvesting and replanting (in the case of annual crops like sugar, cotton, and soy) or pruning and watering (in the case of perennials like tea, coffee, or grapevines). Monoculturing is equal parts caring and killing: pesticides and fertilizers, spades and sickles, irrigating water and combustible fuel (see also Dey 2015).

In the Dooars, as in other Indian plantation districts, national plantation labor laws reinforce the entanglement between monocultures and people. According to those laws, laborers who pluck tea and process it in factories can be paid through a combination of cash wages and in-kind remuneration. As part of their compensation, they and their families are permitted to live in plantation-owned houses and receive biweekly rations of rice and wheat flour, as well as a monthly ration of tea. Upon retirement, workers also receive a gratuity and "provident fund" payments. When not in the fields helping tea bushes to "stand alone," workers must maintain houses, roads, and schools: all of the human support structures that, as with the tea bushes, they do not own. In the tea plantation, "socially necessary unpaid work" and waged work are thus collapsed (J. Moore 2016, 89–90). The laws that codify this blend of in-kind and cash compensation keep marginalized people like the Nepalis and Adivasis who staff Dooars plantations marginally alive but firmly in place.

During her study of Chicana cannery workers in California, one of Patricia Zavella's (1987b) interlocutors described the friendship networks that were engendered through work as a form of "abnormal intimacy."[3] With this phrase, this woman sought to differentiate the process of making friends and relations with people through work from the process of doing so in non-work settings (the "normal"). Intimate relationships generated through acts and spaces of labor might sometimes span beyond the shop floor or the break room, but in this form of intimacy, the worlds of work and life are difficult to disentangle (see Weeks 2011). Intimacy both makes work tolerable and extends its reach. Such sentiments are also key both to resisting capital and to enduring it. Zavella points us here to the fact that work engenders sociality that spans beyond the factory floor. Extending this, work generates relationality among the diverse human and nonhuman participants in work.

On plantations, there are few intimate relations that are not entangled with work; indeed, intimacy can span across species lines. Tsing (2012, 2015) has written extensively about the affective dimensions—the affective differences— between industrial and nonindustrial relations with plants. In a discussion of the affective ecologies of sugar, she draws a comparison between her own research in Kalimantan with swidden cultivators and the work of Sidney W. Mintz (Tsing 2012, 148). In *Worker in the Cane*, Mintz (1960) describes an antag- onistic—even violent—relationship between laborers and the sugarcane they cut on Puerto Rican plantations. Mintz's descriptions of taking down the cane are dramatic. The plant is sharp, dry, and adversarial. Mintz writes,

> Long lines of men stand before the cane like soldiers before an enemy.
> The machetes sweep down and across the stalks, cutting them close to
> the ground. The leaves are lopped off, the stalk cut in halves or thirds and
> dropped behind. . . . The hair of the cane pierces the skin and works its way
> down the neck. The ground is furrowed and makes footing difficult, and the
> soil gives off heat like an oven. The *mayordomo* sits astride a roan mare and
> supervises the field operations. . . . To see him ride past a line of men bent

over and dripping sweat, to hear the sounds of the oxen in the fields behind, the human and animal grunting, and to feel the waves of heat billowing out of the ground and cane evokes images of other times. (1960, 20–22)

But Tsing points out, "Human-cane antagonism is not inherent in the nature of cane plants" (2012, 148). Instead, she describes a different relationship between people in the Indonesian rainforest and the sugarcane they find in swiddens. Unlike in Puerto Rico, cane in Kalimantan is a treat. The cane in which Mintz's informants worked was standardized, industrialized. Sugarcane in Indonesia, by contrast, was a result of Meratus farmers' care for swiddens. Tsing (2012, 148) suggests that plantations "remove the love" between people and plants. In place of that love, "European planters introduced cultivation through coercion" (148).

While we can see loveless relationships between people and plants in mono-cultures, other forms of intimacy are key to how monocultures work. And these intimacies, as Zavella reminds us, are forged through the rhythms of work. Without an understanding of such intimacy, it is difficult to understand the mix of starvation, hope, and plantation abandonment in the Dooars. In monocultures, intimacies keep owners, workers, families, machinery, and tea plants entangled together ecologically, economically, and affectively. As I have argued elsewhere (Besky 2014), a plant's ecology shapes the forms and, more importantly, the meanings of the labor necessary to reproduce it. The ecol-ogy of tea is qualitatively different from that of cane, cotton, or coffee—other well-known colonial plantation crops. Coffee, cane, and cotton are harvested in intense, short-term cycles, at the end of which fields are barren, with nothing but stumps or fruitless bushes remaining.

Tea is not harvested like sugar and, arguably, not harvested at all. Two leaves and a bud of tea must be plucked and pruned by skilled hands, branch by branch, sprig by sprig, and brought to the processing factory to produce the right consistency and taste every day, ten months a year. As a socioecological system and a particular labor process, then, the tea plantation requires labor-ers who not only live in but also care about the landscape. As one of Zavella's friends in the cannery put it, "I'm not exactly in love with my job" (1987b, 99). This affective ambivalence is central to work-based intimacy. The interspecies relations of monocultures, of tea, like those of pork, sugar, and soy, sit some-where between alienation and love (see Blanchette, this volume; Hetherington, this volume; Nading, this volume). In fact, key to an analysis of endurance and exhaustion in contemporary capitalist production may be an attention to the

spaces and acts of alienation as well as spaces and acts of belonging and meaning making. Attention to endurance and exhaustion calls attention to the works
of maintenance that make things, ideas, and relationships persist.

Women plantation workers I know often tell me that the productive lives of
tea bushes parallel those of humans. The capacity of bushes to produce viable
leaves—their capacity to *work*—dissipates over time, just like that of humans.
Tea workers I have interviewed frequently talk about the plantation—including
tea plants—through kinship metaphors. Women tea pluckers described themselves as mothers and grandmothers to tea bushes, and male plantation managers as the paternalistic "uncles" (*kākā*) to plantation women. Plantation labor
was simultaneously affective and oppressive. Women articulated their labors as
care for an industrial agricultural landscape, but they also complained about the
rainy monsoon days, tedious repetition of plucking, and unsympathetic management. They lived and performed this work on a plantation landscape that
was owned and controlled by increasingly austere planters.

Tea pluckers experienced the world not just alongside tea bushes but along
with them. The material condition of tea, whether withered, delicate, dry, or
vibrant, did not just reflect the material conditions of workers' lives. Rather, in
this form of interspecies intimacy, these conditions co-constituted one another.
If tea pluckers felt they were unable to take proper care of their tea bush children and grandchildren, this had direct implications for their ability to care for
their human offspring. Importantly, workers were not able to choose the conditions under which they cared for their plant *or* human children.

Filtered through notions of kinship, the work of monoculture tacks "back
and forth" between "what counts as natural/given, and cultural/created"
(Franklin and McKinnon 2001, 16). Tea workers' intimate relationships with
bushes *exceed* alienation, even if they fall short of care. Over generations, they
have become practiced at balancing intimate engagement with tea as a fellow
occupant of the landscape with detachment from tea as an undifferentiated
bundle of leaves and buds. Detachment and abandonment, however, are different processes.

Sickness, Exhaustion, and Endurance

In January 2016, I was at Stellabarrie, a Dooars plantation whose owners had
shut down operations in September 2015. Citing a lack of profitability, management stopped paying wages and benefits to workers. I visited one Sunday. To

get to the plantation, I had to cross a dry riverbed near the plantation, which filled with picnicking families from the nearby city of Siliguri. That Sunday at Stellabarrie, members of a Siliguri-based bankers' union were handing out food aid, in the form of six-kilogram portions of rice, to the workers who remained on the plantation without pay.

The food distribution took place at the plantation factory. Like most factories, it was surrounded by barbed wire and only accessible via a large gate. The workers were lined up single file outside of the gate with empty shopping bags, waiting to hand in little square, pink paper chits that would allow them to get their scoops of rice. The workers filed in, ten at a time, collected their rice, and exited the gate. A woman at a table inside the gate signed each worker's chit to prevent them from coming back again to collect more rice.

Outside the factory at Stellabarrie where the workers received their rice rations, a doctor had come for the day to examine workers and distribute medicines. The ad hoc diagnoses included tuberculosis, low-grade fever, and cough. Nonsteroidal anti-inflammatory drugs were given out for generalized pain and swelling. Most of the women were anemic. With some stinging irony, they were told to eat more vegetables, and when they could, a little meat.

Stellabarrie had opened and closed several times over the previous few years, and workers with whom I spoke while waiting in line for medicine or food were hopeful that it would reopen by spring. This hope was rooted in both past experience (the plantation did tend to reopen) and in a quirk of Indian plantation law. The land under tea plantations in the Dooars is technically owned by the state of West Bengal and leased out to individual companies. When a tea plantation stops operations, its owners have a fixed amount of time to reopen or risk losing their leasehold. If an owner loses a lease, the fixed capital atop the land (the factory, the bushes, the infrastructure) can be sold to a new buyer, but the land stays with the state, and the new buyer must operate a plantation with it.

To avoid a forced sale, owners can declare their plantations to be "sick." "Sick" status, in this respect, is akin to bankruptcy. The state refrains from repossessing the leasehold, and the original owner has the opportunity to refinance and reopen. Plantations like Stellabarrie, which were chronically open and shut, were "sick" in this sense. Throughout the rash of closures in the Dooars over the last few years, owners reported to journalists and government officials that they fully intended to resume production. Plantation owners and managers maintained that this was only temporary. State and national politicians and bureaucrats as well as local political leaders all looked the other way.

The idea that tea plantations can become sick (and potentially recover) is a powerful one not just for Dooars plantation owners seeking insulation from state sanction but also for workers. What is sick can be healed. The food aid from the bankers was one of many relief actions that took place in the Dooars that winter. The response to stories of starvation from the state of West Bengal and the Indian central government had been lethargic, uncoordinated, or both, in part because of the legal structures that legitimized declarations of plantation "sickness." Relief thus came from Christian groups (many Adivasis in the Dooars have converted to Christianity), local nongovernmental organizations, and other volunteer groups. The distribution of food and medicine at Stellabarrie had all the choreography of a humanitarian intervention, but the mood among the workers was one not of desperation. Rather, it was one of cautious optimism. Food relief was just that: temporary relief. A treatment of symptoms alone.

When I asked plantation owners and managers why sickness was so common, they told me a story of ecological and economic exhaustion. The bushes were old and overworked, the machinery was out of date and in need of repair, and the cost of supporting the full-time workforce was too great. The decision by management at Stellabarrie and other Dooars plantations to stop providing wages and benefits when they did (in September) made sense in the context of the ecology and market for tea. By September, as the monsoon reaches its climax, tea bushes still yield leaves, but the value of this "autumn flush" is much less than the "first flush" and "second flush" harvests that take place between March and June.

Stellabarrie's *garden bau*, a leader among the workforce and an ad-hoc liaison between labor and management, insisted that Stellabarrie *could* operate year-round like any other "healthy" plantation, but its owners had neglected to "invest" in or "repair" the bushes. On these "healthy" plantations, the winter dormant season, between November and February, is a time when workers prune the tea bushes so that fresh young sprigs will grow when the weather turns warm again. Bushes that had grown old and unproductive were not being replaced, as they should have been, with younger, healthier clones. The Tea Board of India had actually initiated a scheme to finance replantation on Dooars plantations, but Stellabarrie could not take advantage of it because management had missed too many wage and pension payments during its "sick" periods.

As the garden bau explained, Stellabarrie's management stopped payment at a crucial time not just in the ecological calendar but in the social calendar as well. On healthy plantations, management is obligated under plantation labor law to pay each worker a cash bonus in October to mark the Hindu festival season. The October bonus is vitally important for social reproduction on plantations. October is a time for visiting family, for decorating and cleaning houses, for discussing marriage prospects, and, of course, for eating and drinking. On healthy plantations, "reinvestment" in tea bushes coincides temporally with reinvestment in household, kin, and social relations. This meant that the September closure at Stellabarrie was particularly painful. At Stellabarrie, workers had already started making arrangements for the festivals, which left them with no cash savings by the time management had stopped paying wages.

The workers in the food line told me that six kilograms of rice was minimal, but still, they took it. "We are old and we can't do anything else," one woman explained. "What can I do? I sit in the house and wait."

While older women "waited" on the plantation, others looked for wage work near the plantation.

"Most of the people are in the river," said the garden bau's wife, who was collecting chits at the gate. The rivers were wider and flatter here in the Dooars than in the steeper hills where most orthodox tea was grown. In the dry winter season, riverbeds become spaces of labor and recreation. Picnickers like those I had seen on my way to Stellabarrie could get to them with ease. So, too, could construction trucks, driven by those looking to harvest the stones left in the bed. Able-bodied tea pluckers—mostly women—could walk to the river and break rocks, working them into neat square piles, ready for shoveling into the beds of the trucks. Rock breaking is not easy, but for many in the Dooars, it is preferable to migrating to Delhi, Bangalore, or Kerala in search of low-wage work. In the big cities, there were little more than domestic service jobs available to Nepalis. Adivasis, on the other hand, did not have the social networks that facilitated migration. In addition, off of the plantations, few people spoke Nepali or Sadri.

There were thus good reasons why workers on sick plantations remained in place, waiting, including the geographical isolation of the Dooars, linguistic and ethnic discrimination, the need to care for aging relatives, and the fact that workers' houses were attached to the plantations. That seasonal sense of hope was also important. The first flush of March was coming, after all. Immobility is built into the monoculture and inflected by a kind of abnormal intimacy:

intergenerational care in a house owned by one's employer and interspecies connections mediated by the seasonality of ritual, rainfall, and temperature.

In the Dooars, plantations were clearly deteriorating, as were people's livelihoods and the bushes themselves, but workers did not leave. Understanding why workers insisted on staying—even on closed or "sick" plantations—requires thinking of monoculture as something more than work *on* land or plants. It is something more than homogeneity and mere alienation. In monocultures, human life, plant life, human work, and plant work are far too intertwined for such an analysis to hold. The plantation was lived *in* and worked *with*, not *on*. Attention to this process of working-in and working-with forces a reevaluation of the common trope of plantations as solely "industrial farms," or even as feudal vestiges.

In monocultures, this immobility and intimacy creates dangers not only for labor but also for capital. The longer Stellabarrie stayed "sick," the less likely it would be that its owner would find a new buyer. In order to take over the lease, any new owner would need to pay the back wages and pension contributions from the date of closure. According to the terms of the lease from the state of West Bengal, the new owner would also have to agree to keep Stellabarrie running as a plantation. To rip up tea bushes and not replant them with more tea would be a violation of the lease agreement.

The lease requirement to keep plantations running *as plantations* exhausts the bushes and the soil, but the promise that a winter's rest will lead to recovery, both of the plants and the people who care for them, allows the monoculture to endure. On healthy plantations, workers were compensated for aiding in this recovery, for doing the work of maintenance. This was the annual work of pruning and manuring. But on sick plantations like Stellabarrie, they were not.

Insofar as they are connected to the stoppages in wages and benefits, starvation and illness among workers on sick plantations are symptoms of an overworked monoculture. The food aid workers at Stellabarrie, the doctors, and even many women's willingness to break rocks were all attempts to *revive* the monoculture, not change it.

Conclusions

According to some activists and even many long-time tea planters, sellers, and brokers I interviewed, chronic plantation sickness was a sign that this entire system should be abandoned. In late 2015, I spoke to a Nepali labor activist who was the child of tea plantation workers in the Dooars. Like others familiar with working conditions in the tea fields, he was cynical about the fact that it was only amid the recent spate of "sickness" that national newspapers had bothered to report about starvation, a lack of medicine, and other forms of exhaustion. To him, these were not new phenomena. The region and its plantations had been "sick" for some time, even if plantation owners had not publicly declared them so. The starvation, the chronic (human) illness, and the chronic loss of profits signaled to him that everyone should just walk away and try something new. Sick bodies and sick landscapes needed to be sacrificed, not saved.

As Povinelli (2011) argues, contemporary critique and politics are hamstrung by the assumption that calls for abandonment—for example, statements like that of the activist—must always be paired with normative, programmatic alternatives. As she notes, however, those who refuse the responsibility to formulate such alternatives are not failing to act politically; they "are acting positively in a social world that is built in such a way that it is unreliable for them whether or not the statement 'not this' immediately produces a 'what then'" (Povinelli 2011, 191). In a way, the activist I interviewed and others like him were saying "not this" to a colonially entrenched system that generated "cheap nature" at the expense of soil, water, and marginalized people (J. Moore 2015). He presented no specific alternative. Indeed, his suggestion that the bodies and land should be sacrificed seems almost absurd: an expression of exhaustion.

Tea in India might seem to have a biological affinity for "standing alone" (Tsing 2015). It has adapted to the high Himalayan foothills as well as the hot, sunny Bengal plains. It has also proved remarkably resistant to drought, pests, and blights. But the term "standing alone" may not be totally appropriate when we consider the spectrum of works and lives that hold the plantation together. That work takes place in the space between exhaustion and endurance, alienation and entanglement.

The legal and vernacular category of the "sick garden" is, like the aging factories, houses, bungalows, and other trappings of the tea plantation, a particular kind of artifact. Just as capitalist ruins contain an immanent potential for new relationships of work and value, these elements have a special capacity for endurance and entrenchment. Taken together, these elements are the sites and structures through which intimacies are forged among labor, capital, and plants. Sickness is becoming something of a norm in the Indian tea industry, especially in those parts of it that are oriented to the production of CTC. In times of sickness, the violence of monoculture is intensified, and its more-than-human relational politics rendered more clearly visible.

Across the Dooars, the monoculture continues, however haltingly, to endure. Capital endures because there is always a little more money to be made from those early spring flushes. Labor endures because home is on the plantation. Those children and grandchildren who have gone off to work in Delhi or Bangalore need a home to return to in October. Ensuring this return requires working through the structures that keep Nepalis and Adivasis tied to the monoculture in the Dooars, rather than overtly opposing those structures. Capital and labor both owe their endurance, in turn, to the tea bushes — old, withered, but after a bit of rest, still able to provide enough cheap tea to keep everyone around. In the monoculture, endurance and exhaustion are thus in constant tension. Attention to exhaustion and endurance provides a view of monoculture as not simply a space of lack or negativity but as a space in which biophysical, symbolic, and technological work are steeped together.

Notes

1. For a discussion of tea plantation work and life in the Dooars and Assam, see P. Chatterjee (2001) and Sharma (2011).

2. The following passages are from the anonymously written *Tea Cultivation* (Calcutta: Military Orphan Press, 1865). The manual was written for tea planters in India, particularly those working in the Dooars and parts of Assam, and was published immediately following the annexation of the Dooars.

3. The "normal" here is outside of or prior to the space of work. This notion of intimacy is useful to think about the relationships that are forged through the process of work that transcend space and kin relations. I shorten this concept to "intimacy" in the remainder of the chapter for clarity.

CHAPTER TWO

The Concentration of Killing

Soy, Labor, and the Long Green Revolution

KREGG HETHERINGTON

In late 2004, a new coalition of activists emerged from eastern Paraguay calling itself the "Front for the Defense of Sovereignty and Life," aiming to counteract what it saw as an existential threat to campesino (peasant) life on the eastern frontier. Small farming communities had been created during the 1960s and 1970s as part of a frontier resettlement of Paraguay's eastern border region on the basis of family labor and state-backed export crops, especially cotton. But these communities had been in economic decline since the 1990s. Now, Brazilian migrants were buying up land in these communities, consolidating it into large soybean fields that could be harvested mechanically. The Front opposed land concentration, and as their name suggests, they were built around a nationalist rejection of Brazilian settlers. But they also opposed soy in general, claiming that soy farms led to widespread killing: deforestation, poisonous contamination, and even murder. This contention they summed up with the slogan "Soy Kills."

I have written about this movement before (Hetherington 2013), arguing that it was very effective at highlighting both social and environmental problems associated with the concentration of agriculture in the region. What interests me here, however, is trying to understand the theory of "life" implied by the statement that soy kills. As Eugene Thacker suggests, the category "life" often stands for something indistinct, the formless subject of threat: "Something is always happening *to* 'life,' as that which is already expressed, already operative, already qualified" (2010, 33). To say that soy kills, therefore, works in part by avoiding the responsibility for qualifying that which is being killed. There are times, of course, when such vagueness is politically necessary, and here it helps establish a moral binary that invites powerful disagreement. But there are other moments when precision around political terms opens up new ways of thinking about a situation and highlights the complexity of a matter of concern. This chapter aims to render more precise the implicit analytics of the defense of life in Paraguay.

In animated discussions about how to tackle the soy boom, two important ambiguities emerged among Front members. The first was the recognition that many of those planting soy were not unlike themselves, often poor and indebted, and that their alliance with an extremely profitable plant was similar to the way campesinos had once experienced the cotton boom of the 1970s and 1980s. The second is that the environmental critique implicit in "soy kills" rebounds on cotton farmers as well. Soy farmers' notorious use of bulldozers to destroy standing forests was the most awesome form of destruction that most campesinos had seen, with huge piles of tropical hardwoods being burned so as to more quickly plant soy. The extensive use of toxic chemicals in soy fields, mixed with international activist outcry about genetic modification (and Monsanto's Roundup Ready soybeans in particular) also made them obvious targets. But campesino activists were always forced to admit that the environmental devastation wrought by soy was related to devastation that they themselves participated in, from cutting forest to using "poison" to cultivate their own crops.

Starting from these conversations, and from many years of research with both soy farmers and their opponents, I want to develop the idea of life in three different ways. The first two are explicit in campesino politics. First, life refers to a specific *way of life*, an ethnic nationalist assemblage that is built through association with cotton. Second, life refers to human labor in the general sense and the feeling that the soy frontier removes laboring bodies from the countryside. To this I want to add a third, speculative layer. I begin from the point that agrarian labor always entails killing. By focusing on agriculture's violence during different moments of Paraguayan frontier expansion, from land grabbing to deforestation, ploughing to pesticide use, the specificity of soy's mode of killing becomes clearer. The relevant question about soy is not whether it kills, but *how* it kills, and what relations of living and killing (or living through killing) it promotes. By following the transition from the campesino way of life with cotton to the way of life of the soy boom, I explore the long Green Revolution as a slow process of intensification of killing and a concentration of labor and death in new forms of property.

A Cotton Way of Life

When people in rural Paraguay talk about "campesino life," the way of life currently being threatened by soy farms, they are referring to a way of life that developed around the cotton boom of the 1960s and 1970s. But the story of modern Paraguay's entanglement with cotton goes back at least to the previous century. Paraguay was all but destroyed in a war with its neighbors between 1865 and 1870. It took over a century for it to rebuild. Cotton was part of the rebuilding effort from the beginning, when Paraguay's most eminent scientist began crop-breeding experiments in the postwar ruins. Moises Santiago Bertoni was an idiosyncratic Swiss botanist who had moved to Paraguay in the 1880s, setting up a secluded research station in the Paraguayan jungle in the hopes of creating an agrarian utopia. He authored hundreds of books and papers, including an exhaustive manual called *El algodón y los algodoneros* (Cotton and the Cotton Growers) (1927a), which also served as a manifesto for the rebirth of Paraguayan agriculture.

Bertoni's idiosyncratic view of agriculture gives us a few keys to understanding what Paraguayans mean by a "way of life." A vitalist ecologist, Bertoni (1927b) understood the nation to be a relationship between the vital forces of specific plants, humans, and territories. Vitalism led him to see life as a metaphysical force, a property of organisms that lay beyond the reach of contemporary science. But his ecological sensibility meant that he continued to see living processes in terms not only of the properties of organisms but also in the relationships between them. A Mendelian rather than a Darwinian evolutionist, he abhorred purity and instead sought to continually improve cultivars, and their relationships to one another, through experimental mixing and companion cropping.

The same sensibility underlies Bertoni's much-criticized anthropological works, especially the epic three-volume *La civilización guaraní* (1922), in which he argued that Paraguay's particular mix of Guarani and Spanish heritage had created an especially noble human race. That story, which would soon be useful to Paraguay's short-lived fascist government, came to underwrite the ethnic nationalism of Paraguayan resurgence in the second half of the twentieth century. Into the *character* of the mestizo campesinos he saw around him, Bertoni wrote many of the virtues he had found lacking in Europe: physical stamina, industriousness, self-sufficiency, environmental knowledge, and Christian family values. Like José Vasconcelos in Mexico, who relied heavily on the vigor of Mexican corn to explain his racial theory (Hartigan 2013, 383), Bertoni freely compared human and plant genetic improvement in his story about Paraguayan vitality. But his view of the mestizo was not merely *analogous* to hybrid plants.

Rather, Bertoni was interested in how the character of certain plants worked with the racial characteristics of those who plant them.

In this, there was no more apt plant for Paraguayans to cultivate than cotton. "The life of cotton," Bertoni wrote,

> cannot be maintained with scattered and adventurous elements, with an expensive or demanding workforce, with flighty or mercenary personnel, and it cannot abide populations with disorganized customs, restless character or industrial habits. It is a family cultivar, and demands a family. It is a democratic plant, autonomous, requiring personal initiative, especially for agrarian colonists who know how much their independence is worth. Paraguay can, and must, be a cotton-producing nation. (1927a, 6)[1]

In other words, to Bertoni, the improvement of the Paraguayan nation was a multispecies project that relied on the relationship between human and plant varieties. Indeed, the ambiguity of the Spanish word *algodonero*, or cotton producer, captures the perfect fusion of his ideas: algodonero refers to both the plant and the farmer. What I've translated above as "cotton-producing nation" really means simultaneously "a nation of cotton farmers" and "a nation of cotton plants." For Bertoni, the cotton nation needed to be underwritten by the exceptional character of campesinos and by scientific investment in improved varieties of cotton.[2]

Bertoni's vision of national development was stalled by the world wars and by almost two decades of national political turmoil. But in the 1950s, when General Alfredo Stroessner took power and began to consolidate the state around a militaristic agrarian development plan, Bertoni's cotton nation began to take shape. There were two prongs to this emergence. The first was the consolidation of agricultural research. Starting in the late 1940s, and with considerable backing from the US Agency for International Development, the Inter-American Development Bank, and the Rockefeller Foundation, Paraguay's public investment in agriculture grew astronomically. This included both the Ministry of Agriculture's extension programs and semi-autonomous crop research stations that generated the best genetic materials and techniques for smallholders. One of the star scientists to emerge from this system was Hernando Bertoni, grandson of Moises, whom Stroessner eventually appointed minister of agriculture. Under Hernando's leadership, the Green Revolution transformed Paraguayan agriculture into a centralized cotton-exporting powerhouse that by the 1970s had the highest GDP growth in the region.

The second part of Paraguay's Green Revolution was land reform through colonization. At the beginning of the 1950s, Paraguay's rural population remained concentrated in the central departments around Asunción, a fact that hampered agricultural production and made it difficult for the government to assert its control over national territory. Stroessner sought to expand the population into the eastern forest (and toward, in particular, the Brazilian border), and in so doing, spur agricultural production. The eastward expansion focused on rewarding friends of the dictator with large ranches and campesinos with ten-hectare lots that would be planted with a mix of subsistence and export crops. For smallholders, colonization proceeded through the reorganization of the relationship between campesino labor and land. Following a standard Lockean script for imagining frontier property, campesinos were invited to find "unused" land to work. Any changes that they made to the land, from clearing forest to building structures, were called "improvements," which served as the basis for a laborer's ownership claims and eventually their recognition by the state, in the form of official property titles, as economic subjects (Hetherington 2011).

One of the core images of the land reform captures the aspirational structure of agrarian colonization in this period. It begins with the landscape depicted as a blank space, the proverbial "desert" of the Latin American frontier. The catalyst for development of this space was usually thought of as man and soil, but the picture can also be read as being about the relationship between a man and a plant, from which follow a heteronormative family, a house, a cultivated field, and legal papers. The final goal was not just economic growth but a "peaceful agrarian revolution" and the "founding of a new Paraguay" (Frutos 1976, 5–6). In other words, campesino labor was not only about producing value from the land but was also part of a republican project.

None of this is to say that land reform and the cotton frontier were a smooth process. Reform almost immediately led to an organized movement that blamed first Stroessner and then succeeding governments for not delivering fully on its redistributive promises. Initially, the concessions of large swaths of land to government cronies was seen as hypocritical, and the frontier included periods of widespread land occupations. From the 1970s onward, campesino occupiers were met with often brutal repression. But however unevenly it may have played out, it is this community of pioneer aspirations that created the campesino way of life as many rural people understood it at the end of the

Figure 2.1. Cover of *De la reforma agraria al bienestar rural* by Juan Manuel Frutos (1976).

twentieth century, with a tense but workable relationship between the state, campesino movements, and the ranchers with whom they shared the landscape. By the turn of the century, as soy expanded and land reform all but halted, many campesino dissidents began to view soy, rather than the state or ranchers, as the true threat to their way of life.

Killing a Way of Life

Soy was not the only force working against campesinos. The cotton crop had spiraled into crisis in the 1980s, first because of depressed global prices and second because of insect pressure. In the early 2000s, I lived near a community I'll call 3 de Noviembre that had been founded in the late 1970s as a stepping-stone to a better life. By the time I came to know it, though, the aspirational qualities of life in these types of towns were dwindling. After the coup that deposed Stroessner in 1989, and in a climate of neoliberal restructuring, crop research and extension were defunded or outsourced to nongovernmental organizations, and the state ceased to play a central role in bolstering the cotton frontier. Towns like 3 de Noviembre were ripe for takeover.

Soy had actually arrived in Paraguay in the late 1960s and established itself in departments between the campesino area and the Brazilian border to the east, carried primarily by Brazilian immigrants. Many of these immigrants, or "Brasiguaios," had themselves been displaced by Brazil's own "March to the West" and created pioneer colonies on Paraguay's far eastern frontier that looked not unlike campesino colonies (Souchaud 2002; Blanc 2015). Many even began by planting cotton but turned to soy farming as the resources (primarily technical and infrastructural) became available on the Brazilian side of the border. In the 1990s, soy farms boomed and expanded, forests inevitably disappeared, and soon Brazilian migrants began to buy up campesino plots in places like 3 de Noviembre (see Glaucer 2009; Elgert 2016). Paraguayan campesinos remained shut out from this wave, culturally excluded by the soy industry run almost entirely in Portuguese and without the capital or credit to invest in the machinery required for planting soy.

There are striking similarities between cotton and soy. Both are frontier boom crops that for different reasons became associated with a particular ethnic population. Campesinos once called cotton "white gold," and Brazilians now call soy "green gold." And it's easy to find soy farmers, and many of their boosters, who will credit their European descent (the majority of Brasiguaios are second- or third-generation German immigrants) for their agricultural success. What Bertoni saw as the greatness of Paraguayan mestizos gets turned on its head in this story, and campesinos are seen as too lazy and primitive for the complex financial and technological demands of soy farming. Soy, by contrast, becomes the crop of whiteness, modernity, and entrepreneurialism.

And yet there are important differences in emphasis about how soy farmers narrate their relationship to soy. In contrast to the land reform propaganda depicted above, the soy frontier has never been organized by a national program, and so there is no official picture or motto that declares its logic. Instead, the visual emblems of Paraguayan soy are strewn across billboards all over eastern Paraguay, in advertisements, and in trade journals. Corporate logos, from the multinational giants Archer Daniels Midland and Syngenta to local financing and chemical firms like Agrotec, Microquímica, and Agrofértil, almost uniformly include a broad, green leaf. In promotional materials, these are generally accompanied by photographs of impressive, massive machinery, pristine green fields stretching to the horizon, or conical beige mountains of harvested soybeans. People rarely appear in these pictures, and when they do, they are often hidden behind the windows of tractors or admiring a load of beans falling from the harvester's chute into a truck.

Rather than human laborers, the beans themselves have become the protagonists of development, in a relationship with international capital, biochemical expertise, and imported machinery. If one can call this a "way of life," it is also both human and plant, but it is the plant that takes center stage, and human relations are configured through it. The most extreme version of this was Syngenta's notorious ad for its seed that featured a map of South America without national borders but with a gigantic green blob covering much of Paraguay, Argentina, Brazil, Bolivia, and Uruguay and captioned "United Republic of Soy."

It's not surprising that a lot of what goes under the heading of "soy kills" has an ethno-nationalist edge. Anti-Brazilian sentiment runs strong in both campesino communities and Paraguayan academic work (Albuquerque 2005; Wagner 1990), to the point that it is quite common to hear the claim that soybeans are merely the latest invader in a war that has been going on between Paraguay and Brazil since 1865. Soy kills the same way that the Brazilian army kills, to make room for Brazilians. But all of this is infused by a second uncanny sense in which soy seems to displace not just a human *way of life* but humans in general. Between the two frontiers we also have a move in the conception of life from the particular human to the generic human, and from the generic plant to the particular plant. Where the wealth of the nation once rested on the proclivities of the Paraguayan "man," the wealth of people, and of humanity, now rests on the power of soy. And it is a fair question, then, whether the newest wave of agriculture is merely a move away from a Cold War commitment to including human laborers in the agrarian economy.

Killing Labor

If in some contexts "life" primarily referred to this specifically campesino way of life, the notion of killing that plays out on the Paraguayan countryside also invokes a more abstract relationship between labor and life. When I first started asking people in 3 de Noviembre what they objected to in soy farms, many of them told me that soy farms didn't give *changas*, or day wages, the way that cotton farmers and ranchers had once done. The importance of the changa to the cotton way of life underscores the fact that even on the cotton frontier, labor was always more than a simple Lockean way of appropriating land. In contrast to the work that family members do on their own plot, the changa is a form of day labor carried out on someone else's farm for a set wage. It is a convenient way of distributing labor around the very uneven rhythms of agrarian life and a way of dealing with inequality, not just between cotton farmers but between peasants and ranchers, an older political elite who manage their status by giving changas as patronage. In her chapter in this volume, Sarah Besky argues that tea pickers expect plantation owners to take care of them and that this complicates an otherwise straightforwardly exploitative relationship. The moral economy of the changa is similar, only it operates between formally independent owners and workers. But since soy farmers were not part of this same moral economy, they could not be counted on to reliably give changas.

I did not carry out sustained research on local labor markets in Paraguay, but I did record one important indicator of this shift. Over the two years that I kept going back to 3 de Noviembre, as larger portions of the surrounding landscape were planted in soy, changas did indeed dry up rapidly. In the spring of 2004 in 3 de Noviembre, young men were still making 15,000 Gs (just under US$4) for a changa, or one day of manual labor. By 2006, *changueros* in the district were only making 8,000 Gs per day. There's a classic argument in agricultural economics that smaller farms are more productive (in terms of value produced per hectare) and are associated with more equitable distribution of wealth and income (Berry and Cline 1979; Carter 1984). In his 2010 study of rural labor in Paraguay, Toledo (2010, 90) found that despite far poorer infrastructure and markets for small farmers than large ones, the smallest farms still produced thirteen times as much product per hectare as the largest, while employing almost twenty times as many laborers. Correspondingly, as soy has moved westward from the Brazilian border, colonizing one district after another, the population has declined in those districts, and the levels of income inequality increased (Elgert 2016; Berry 2010; Fogel 2005; Fogel and Riquelme 2005; Rojas Villagra 2015).

Much of scholarly criticism of soy fits into a larger debate about "land grabbing" that is waged on the relationship between labor markets and farm size. One of the things that's tempting about this argument, and one of the reasons that it dominates so much of the scholarly debate, is that by translating the effect of the soy frontier into a mathematical relationship between property size and labor units, analysts can remove from the equation the ethno-nationalist taint of the argument explored above. It also removes the plants from the discussion, or at least removes any need to talk about their specificity in regard to ways of living. Here "soy kills" comes to mean "soy diminishes human labor in the abstract" by concentrating land and replacing laborers with machines.

This transformation of living fields into soy fields is often referred to as "mechanization," since so many changuero jobs are being taken away by machines. And in this curious way, we can see that the argument that what soy kills is "labor" in the abstract is related to another strand of vitalist thought, which sees human labor as a specific kind of life force that produces value. That tradition begins with Adam Smith's understanding of productive labor as a metaphysical substance that "fixes and realizes itself in some particular subject or vendible commodity, which lasts for some time at least after that labor is past" (quoted in Schabas 2003, 270). This notion that labor is an abstract living force is famously taken up by Karl Marx, who states that capital (including machinery) is "dead labour, that, vampire-like, only lives by sucking living labour, and lives the more, the more labour it sucks" (1976, 342).

But the productive ambiguity of the word "life" here also points to some of the stickiness of thinking about the problem of soy as merely a reduction of the amount of rural labor kicking around. As Thacker (2007, 314) points out, this tradition always struggled with the difficulty of separating human labor force from a larger conception of "life force" that was not specifically human. Marx famously dealt with this problem using the wobbly Aristotelian distinction between men and bees,[3] which returned with a vengeance in the 1970s just as people began to worry about deindustrialization in the United States. To cite only the most obvious example, Harry Braverman's 1974 opus on mechanization, *Labor and Monopoly Capital: The Degradation of Work in the Twentieth Century*, devotes its entire first chapter to elaborating on Marx's dubious defamation of bees. Braverman's insistence on this humanist point is a product of the moment in which he is writing, after World War II, when so many social scientists worked to remove vitalist traces from their thinking by enforcing a strict separation between the biological and the political via a distinction between the animal and the human (Esposito 2008). This mirrors the move in agricultural science, from a model of vitalist interrelations between humans and other living beings to one in which the phytological and the sociological were handled by completely separate institutions, answering to distinct demands. Bertoni and people like him became an embarrassment, to be replaced by two separate forms of expertise: crop scientists who could generate better plants, on the one hand, and sociologists and economists who could figure out how to better distribute human labor, on the other.

But this humanist take on labor as life force only gets us so far in understanding why cotton and cattle, but not soy, give changas. Campesinos will point out that the changa is not just a unit of abstract labor but a very particular way of relating to work, which they continue to value highly. Even though it pays poorly, and tends to be menial, campesinos like changas as a source of quick income that allows them to remain independent. Many will say that it is preferable to be a changuero than to be a *tembiguai*, dependent on a boss.[4] What makes the changa so desirable is that ultimately, a changuero retains his right to refuse work. If it rains, or if your friend wants to go fishing, or if a family member is in need, you don't need to changa that day. Changa sometimes even works as an explicit form of redistribution. So, for instance, a particularly important kind of changa is simply "cleaning," an activity that included all sorts of low-grade fieldwork, like land clearing, hoeing, hand weeding, and brush burning. In 3 de Noviembre, people in dire need of income go door-to-door offering to clean the fields of wealthier neighbors. The changa thus eschews the *duty* to work for a living in favor of a *right* to work that allows one to live.

Against the backdrop of other labor histories, most notably US slavery, the smallholder-changa system seems like a surprising institutional pairing for cotton. But this autonomous work ethic functions well in a landscape of low-intensity farming with an oversupply of active laborers. As Bertoni might have said, if cotton is a democratic plant that fosters autonomy, then changa is its most appropriate form of labor. And indeed, even though campesinos use the term *precario* to describe the situation of poverty and uncertainty within which they live, most are also loath to give up the freedom it offers them to live a life that they deem worth living. The same attitude toward work is what led to the survival of various kinds of illegal share-cropping arrangements, whereby campesinos would rather live on someone else's land with a promise of sharing a portion of their crop than engage in a contractual wage relationship or take on heavy debt for property. Following the autonomist Marxist tradition, we might say that campesinos have a propensity for resisting both private property *and* the organization of work (Weeks 2011, 96–101). Well aware of the way work is organized on modern soy farms, many campesinos are explicit that it's not the life for them and will wryly say that one of the ways that soy kills is by working Brazilians to death.

From the point of view of soy farmers, the changa ethic is simply laziness, often expressed in racist terms. But one can see where their frustration comes from, as the high stakes of large-scale farming made uncertainties around labor very difficult to sustain as soy farms grew and farmers became more dependent on this unpliable labor force. These tensions become most clear in moments when a sudden flush of weeds or insect attack put a crop at risk, and it is difficult to round up the manual labor necessary to hoe or spray down the intruders. Every soy farmer I've talked to about the decrease in employment on the soy frontier insists that their disagreements with changueros is one of the reasons why they eventually replaced most of their hired hands with expensive machinery. What all of this suggests is that the killing of a way of life, and the killing of jobs, are in fact related to a third kind of killing, the killing of pests.

Pests and the Acquisition of Killing

Agriculture is all about intimate multispecies relationships, relationships that necessarily involve killing. Harvesting, slaughtering, burning, cutting, ploughing, poisoning, shooting, clearing, trapping, weeding, culling, and selecting are all forms of killing. Even growing a leguminous cover crop between rows of organic corn and squash is a way of choking out undesirable organisms that are constantly trying to take over a garden. As Donna Haraway (2008) might put it, agrarian work is all about figuring out how to make certain combinations of companion species flourish at the expense of others. By extension, the kind of specialization that made it possible to scale up modern agriculture was also about finding new and more efficient ways of killing, and distributing these as specialized forms of work, and figuring out how many of them could be automated.[5]

Consider crop breeding. Since the late 1940s, when the Green Revolution first came to Paraguay, government crop scientists have worked at a network of research stations, recently formalized as the Instituto Paraguayo de Tecnología Agraria (IPTA). This is where Hernando Bertoni got his start as a wheat breeder, and where other researchers produced the cotton varieties that soon dominated eastern Paraguay. One of these stations, at Capitán Miranda, is devoted primarily to grains and legumes, creating the wheat, corn, and bean varieties that fueled Paraguay's internal markets for staples. By the time I began doing research there in 2010, cotton breeding had disappeared, and the largest portion of research was now devoted to soybeans, but the breeding process looked much the same as it had a few generations back.

In Capitán Miranda, a young breeder named Natalia Sanabria showed me her greenhouse, where she worked alone under the guidance of the station's most experienced soy breeder, nurturing beans to tolerate rust, a fungal infection that had been the scourge of Paraguayan soy farmers since the early 2000s. For close to eight years, the team she was part of had been selecting beans that were both tolerant to rust and could be backcrossed with Roundup Ready varieties to produce glyphosate resistance. "You have to be a special kind of person to do this work," she told me, "because it is so arduous and so long, and you're not guaranteed of anything at the end of it all; you need to love spending your time with the plants."[6] Most IPTA scientists I met, from the plant breeders to the soil scientists and entomologists, approached their work with deep care and commitment to their plants and microbes. That care mostly involves killing. Natalia's careful work involves planting rows of forty soy plants, coaxing each of them to the point where their traits can be observed, and then killing thirty-eight of them before allowing the remaining two to reproduce. Three years later, when IPTA would launch a new variety of rust-resistant seed called "SojaPar 19," it would be a result of these years of careful killing. Indeed, so many of the commodities of agrarian production, from seed patents to containers full of beans, are made possible by legal regimes that allow one to take ownership over something living on the basis of having killed many things around it.

If plant breeding and cultivation is killing as part of care, other aspects of the agrarian frontier are far less subtle in their violence. As far back as the 1970s, a handful of ecologists and anthropologists raised the alarm about just how much killing was actually going on for cotton. Peter D. Richards (2011) estimates that as late as 1960, there were still seventy-three thousand square kilometers of standing forest in Paraguay, but that over the subsequent four decades, 75 percent of that was cut down by ranchers, grain farmers on the Brazilian border, and cotton smallholders (with another 13 percent disappearing in the next ten years). More notoriously, but no less surprisingly, in 1973, German anthropologist Mark Münzel (1973) reported the genocide of Aché hunter-gatherers going on in the forest. Münzel's account, and several of those that followed (Arens 1976), were familiar compilations of atrocity, from rampant disease to Indian hunting, trophy killing, and an active market in Aché slaves. The Aché unquestionably suffered more than any other Indigenous group in the eastern forests, but their plight was a symptom of a larger displacement of forest dwellers by cash croppers backed by the Paraguayan state (e.g., Fogel 1998).

As the cotton frontier advanced, it required new forms of organized killing. When boll weevils appeared in Paraguay in the early 1990s, it brought forth a national response, importing potent pesticides such as parathion, which could then be distributed on credit through the agrarian development bank and increasingly complicated patron-client networks of the ruling party. There are many reasons why cotton declined throughout the nineties, but one of the key ones was the ability of boll weevils to kill cotton plants faster than farmers could kill them. The diminishing returns on the harvest, and the increasing cost of toxic inputs, created unmanageable debt for both farmers and the state's agrarian bank, and credit became one of the key axes of campesino protest and dissatisfaction with the terms of the agrarian reform.

We can now appreciate a bit more what soy finds so propitious about the old cotton frontier. It is neither a pristine environment nor a place of thriving communities, but rather a blasted landscape and a scene of carnage of both humans and nonhumans (Tsing 2015). As Besky (this volume) points out, the labor of making monocrops like soy fields thrive involves suppressing other living entanglements. In this light, we can revisit the way that frontier property transactions built on the notion of "improvement" made to the land through campesino labor, the first of which was always deforestation. Implicitly, campesino colonies also made other improvements of this sort, such as clearing out Indigenous people who might now be protected by international conventions or hunting the jaguars that some environmental agency might now seek to conserve. Generally, campesinos gave frontier land value by killing off several of the vital possibilities that the land might have held only two generations back. In other words, just as seed patents create property from the killing of less-viable plants, land titles are in part the commodification of a history of killing in a particular place. As the new frontier wave entered, and soy killed off the inhabitants of Paraguay's eastern landscape in new ways, it was not as a completely new or unprecedented displacement of the vital labor that had existed there before but rather an appropriation and concentration of past forms of killing in the project of new, more efficient techniques for generating property from death.

This brings us back to the changa, and the euphemistically named "cleaning" so crucial to the survival of cotton colonies. The machines that replaced changueros are almost all large-scale implements of death: the bulldozer (for land clearing), the plough, the herbicide sprayer, and the harvester, with which one person could kill off much of the biomass in a relatively large area. And it's instructive to look at the latest of these technologies, the Roundup Ready soybean. As late as 1999, most soy farms still needed a steady supply of changueros who could be drawn from nearby campesino communities. This was because farmers could mechanize land clearing, planting, and harvesting but still needed people on hand to deal with weeds that sprung up after germination and before harvesting. A hundred-hectare field of soy could provide work of this sort for at least a dozen men on a fairly regular basis, putting money back into circulation in the campesino communities nearby. What finally did in the changa was genetic modification.

Glyphosate has been a common product in agriculture since the 1960s, when it was introduced by Monsanto under the name "Roundup." A broad-spectrum weed killer, glyphosate inhibits the processing of amino acids, killing almost all plants affected in a few days. If there's one thing for which Roundup is extremely effective, it is "cleaning" a field. In the 1980s, a single pass with the sprayer was sufficient to prepare the terrain for seeding. Campesinos even used it periodically to clean particularly stubborn weeds before planting a garden, referring to it affectionately as "*mata-todo*," or "kills-everything." The great advancement, which became the darling of the biotech industry and saved Monsanto from patent-expiration obsolescence, was the engineering of soybeans that could resist glyphosate (Robin 2010).

Here, then, is what really lies behind the images of clean, wealth-producing soy plants, with so few humans around them, so omnipresent in the promotional material of the soy frontier. The life of Roundup Ready soy derives from its resilience (and therefore self-selection) in the midst of industrial killing — Roundup Ready's internalization of the relationship between killing and survival that had once been resolved with changas. Once changueros are no longer needed, the small farms in the area become vulnerable and can more easily be bought. Access to one technology of killing (Roundup Ready) gives access to another (the neighbor's property title). The expansive soy monocrops of eastern Paraguay are made possible by a steady concentration of different forms of killing.

If there was any doubt of the importance of this for anti-soy campesinos, consider one of the solutions to local conflicts between soy farmers and their neighbors, which I first saw happening in 3 de Noviembre. The previous year, following the pattern of soy incursion into cotton colonies, three small farms of ten to sixteen hectares had been bought by Brasiguaio settlers from a nearby town. Campesinos had immediately begun to complain that these "foreigners" were making their lives unlivable by spraying glyphosate from tractors, leading to everything from an unpleasant smell to severe health problems in the community. The solution to this (which two of the soy farmers agreed to) was to ask them to stop spraying their fields with tractors and instead to hire their neighbors to spray using backpacks (as they did for cotton). The solution didn't last long. Whether because of increased cost or because of an unmanageable labor relationship, the soy farmers gave up after a year. Of the three farmers targeted by protests, one replaced his soy with corn, another canceled his lease and left town, and the third doggedly remained, ultimately leading to a full-blown takeover of the community by soybeans some four years later. What's interesting here is how the fragile resolution clarifies the stakes of what makes soy so powerful: for many, soy could stay so long as campesinos were invited to participate, to thrive even, in the labor of killing that soy required.

But this of course was not the most common fate of campesinos. Between 2005 and 2009, many families in 3 de Noviembre and nearby towns were selling their properties off to soy farmers in different ways. In many places, legal restrictions on land transfers made it difficult to buy land outright without a long period of filing paperwork. One solution to all of these problems was to enter into an informal purchase agreement whereby campesinos would be paid a wage to further improve their own land so that it could then be sold. The improvements included completing the process of titling the property to make it easier to transfer ownership rights and land clearing for future soy plantations. To the extent that campesinos were able to resist the imposition of labor discipline, they eventually gave in to the imposition of private property as a way of valuing and selling their own labor of killing.

In other words, unemployed ex-cotton growers were being paid to remove what was left of their farms, particularly the shade trees, woodlots, orange groves, and old, charred stumps that had always been part of the small cotton farms. Sometimes called "*limpieza*" (cleaning), "*destronco*" (detrunking), or "*mecanización*" (mechanization), to soy farmers, of course, this process was just the final phase in the clearing of forest for agriculture, the finalization of a process that campesinos had already mostly accomplished. To anti-soy activists, however, it was something much more sinister. The symbolic significance of being paid to rip out the plants that they themselves had planted so that they could later be replaced by soybeans was lost on no one. During one of these periods of movement along the road to 3 de Noviembre, in June 2006, my friend Antonio Galeano pointed out a neighbor who had recently finished his limpieza by knocking down his own house and hauling it away on a truck. "They're paying him to kill himself," he said. At the end of the process, the legal titling of the improvements was about acquiring the rights to sell the product of one's own self-killing.

Conclusions

One of the things for which Moises Bertoni is famous in Paraguay is his position against the use of fire for clearing cropland. Bertoni had no trouble with
killing things. A tireless experimenter and plant breeder, committed to the
modernization of export agriculture, at no point in his writings does he argue
that one shouldn't kill an insect that is eating one's crops. But in an essay that
was far ahead of its time, Bertoni (1926) argued that one should avoid the kind
of indiscriminate killing wrought by brush fires, even though these were an easy
and efficient way of clearing land. The long-term effects of such indiscriminate
killing outweighed any of fire's short-term benefits. The paper offers a series
of alternatives to the use of fire, including the most radical, "planting without
cleaning" (1926, 8), a crop system that would today be called "permaculture,"
which focuses on the companionship between crops, trees, soils, and debris in
creating intensively farmed plots with a minimum of killing.

Bertoni saw fire-free farming as an ethical, ecological, and economic imperative, and he was convinced that posterity would agree with him:

> You can be sure, in the not-too-distant future, no one will clear their
> land with fire, no matter what the opposition to this may be at the outset.
> Nobody should harbor the least doubt about this: our descendants will not
> only stop committing this barbarity of burning fields, but they will have
> difficulty explaining how their ancestors could have believed for so long
> that it was the most rational thing to do. (1926, 11)

Fire is still very much present in Paraguayan agriculture. But glyphosate, and
the beans that require it, is surely the modern equivalent of Bertoni's fire. As
Jake Kosek points out in his contribution to this volume, the widespread use of
glyphosate in the United States does far more than reduce plant life; it degrades
the web of relationships that normally keep any living space vibrant.

Glyphosate, like fire, is never completely successful at dominating nature in
the way it is intended to (Pollan 2001, chap. 4). By 2017, Paraguayan soy fields
had become so tangled with glyphosate-resistant weeds that many farmers were
having to scramble to find changueros to clean their fields again. None of this is
necessarily linear, and one can imagine all sorts of ways in which the soy economy might collapse or become something else. But just as the image of citizenship propelled the cotton boom, here it is the aspiration to concentrating the

ownership of killing that propels soy. In answer to glyphosate-resistant weeds, crop breeders at Capitán Miranda told me they were looking forward to the development of beans that resist far stronger herbicides such as 2,4-D and Paraquat. The logic of the long Green Revolution seems to lead toward increasing the efficiency and extension of indiscriminate killing.

The argument that "soy kills" is only useful so long as we push it beyond a purist ethic involving life and death. But if a way of life is always a way of killing, then the campesino movement behooves us to think about how killing is distributed. We can easily look back on the cotton frontier with horror. In its early days, the Green Revolution was no less bent on violence, no less committed to control, than is so much of the soy frontier. The difference is one of concentration, of the aspirational monopoly on destruction that soy makes possible. Quaint as it seems now, the quality of smallholder cotton farming that Bertoni called "democratic" shines all the brighter in relation to the advancing field of beans. If soy and glyphosate were one response to cotton's violence, another might be to reclaim agricultural democracy, in which varieties of plants, insects, and people have to negotiate ways of living and killing together.

Notes

1. Unless otherwise indicated, all translations from Spanish are by the author.

2. This is one reason I've chosen not to adopt Stefan Helmreich's (2009, 6) otherwise useful distinction between "life forms" (organisms) and "forms of life" (ways of organizing human communities). In doing so, Helmreich points to the hazy distinction between these two notions of life while simultaneously upholding a binary. But Bertoni's cotton nation is a lively assemblage that doesn't easily resolve this way, since the human life projects are as much in the plant as the plants are in the human communities. (See also Hartigan's chapter in this volume for a discussion.)

3. This passage is a favorite target for posthumanist critique, which I won't rehearse here (see Ingold 1983; Kosek 2010; T. Mitchell 2002).

4. "Tembiguai" literally means someone who is defined by their work, but it has a long pejorative history and was associated in the nineteenth century with slavery (Huner, forthcoming).

5. Alex Blanchette's contribution to this volume provides an extreme example of this.

6. "Natalia Sanabria" (pseudonym), personal communication, January 2012.

CHAPTER THREE

Making Monotony

Bedsores and Other Signs of an Overworked Hog

ALEX BLANCHETTE

Skin Workers, or the Politics of Pig "Boredom"

Consider the figure of the skin worker. Or, consider the fact that there is now a single person on many American factory farms who spends a few hours each day tending primarily to the porcine epidermis. This person sprays thousands of spurts of iodine onto hogs' legs, bellies, and shoulders in an effort to maintain the rubbed-raw skin of confined animals. The skin worker is a recently instituted position and mode of labor in agribusiness that embodies how this chapter thinks about late capitalist animal productivity. As industrialization intensifies over time, emergent dimensions of pigs are gradually becoming subject to (and subjects of) specialized forms of work. This chapter ethnographically describes how corporations' efforts to make hogs perform monotonous work in service of uniform meat—repetitively performing single tasks in an identical manner, such as gestating or emitting saliva pheromones—requires the creation of human farmworkers who can act as prosthetic supports for other facets of porcine biology.

The pages that follow thus provide snapshots of people laboring on pig skins, muscles, perception, and hormones to critically appraise how agricultural engineers are manifesting hogs that monotonously work and yet require more (human) work. This approach to industrial animality is inspired by the feminist critic Kathi Weeks's (2011) claim that a key political project of the present is to defend social life and aspirations for liberation against modern work ethics. She argues that we should not merely advocate for better work but also politicize and denaturalize the supreme cultural value accorded to human labor. The project of reducing work, however, both in terms of decreasing the number of hours dedicated to labor and cutting down the social value of work, is troubled by these hogs' very existence. For the logic of factory farms is not (just) an industrial narrative of deskilled and exploited manual laborers who must work to live. It is instead one of a species' capacity to survive figuratively "deskilled" to the point that the maintenance of hogs' skin cells and muscle fibers requires human labor. The industrial pig more than reflects late capitalist cultural values—it is a literal embodiment of how life is becoming work, and work is becoming life. The skin worker's knowledge of bedsores is a sign of an *overworked* hog: a product of labor that necessitates increasingly intense forms of work to be sustained.

The person who tended to be the skin worker on Sow #6, the 2,500-head Great Plains corporate breeding farm where I was employed in 2010, was named Luis.[1] I did not get to know him well. While the rest of us were assigned in small groups to artificially inseminate sows, he worked alone with a plastic dollar-store spray bottle of iodine dangling from his coverall pockets. Much of his morning was consumed by studying sows' bodies for rashes, cuts, and especially nascent bedsores. It was not an easy job. When I was once assigned to the task, because Luis was absent, it was difficult to spot even an acute injury in the low-lit barn. I could only notice the fading rusty iodine splotch of the previous day's effort to ward off infection and continue the treatment that a more skilled skin worker such as Luis had already started. Luis, conversely, seemed to be able to spot even nascent injuries, knowing where to look on the contours of any given animal's body—and there are thousands in every barn, all slightly different—to see how a sow's bones might be repetitively pressing against the concrete floor, the bars of her gestation crate, or her protruding water spout. He could detect subtle changes in skin tone that indicate an incipient injury or an odd tussle of fur that suggested repetitive contact. His position in the barns was poorly paid, about $9.50 per hour in 2010. It was the sort of position described by management as an entry-level "manual" job that anyone can do. But Luis had acquired profound knowledge about how the porcine epidermis manifests in conditions of confinement. Moreover, Luis's capacities to support the porcine skin organ, in his ways of looking and gestures, appear necessary to maintain the project of American meat as we know it.

Thinking about industrial breeding farms today, two images leap most prominently to my mind. The first is the frightening sense of calm despite the dense concentration of bodies. The barns are often dead silent. Outside of their loud shrieks in anticipation of morning feeding time, thousands of four-hundred-pound sows tend to just lie on their sides in gestation crates. For most of the day, breeding animals appear to be inactive as they sprawl on the concrete slat floor. The second is the bedsores that can result from this lethargy. The shoulders and rumps of confined sows are pockmarked with scars, abscesses, and the dull red scabs of pressure ulcers. These bedsores elicited much commentary. A couple of co-workers explained their aversion to eating pork by pointing to these injuries of apparent inactivity. Moreover, in an environment teeming with fecal dust and bacteria, the red wound of a bedsore (along with the constant skin-penetrating drug injections) carries a risk of infection. As agricultural science researchers examine the systemic pain, along with death losses due to sow lameness and culling, that can emerge from these injuries (see Hostetler 2012; Rioja-Lang et al. 2018), the lethargy of confined animals is beginning to emerge as an acknowledged production problem.[2]

Prominent critics have also been launching important efforts to politicize the inactivity of industrial animals. The Humane Society of the United States (HSUS) (2010), for instance, released an award-winning undercover exposé of Smithfield Foods, the world's largest manufacturer of pork, that centered on the tedious boredom of confined pigs. While exposé videos usually feature scenes of enraged workers harming pigs, this one was notable for its near absence of human bodies. It depicted lethargic sows locked in crates, red wounds pockmarking their bodies. The video frames stereotypic behaviors, such as abnormal chewing on bars, and inactivity as key injuries of porcine modernity, indexes of a violent stasis. Similarly, a project from activist video game designers in the Netherlands, *Pig Chase* (Alfrink, Van Peer, and Lagerweij 2012), proposes to install web-connected screen projections in that country's confinement barns. The concept is that users will interact with industrial hogs using smartphones, controlling projected lasers with their fingers to earn points if they attract the pigs' attention. Their idea is to crowdsource game players from all over the world to engage in the (unpaid) labor of mentally stimulating Dutch pigs toward play—perhaps, at its best, even creating digitally mediated forms of interspecies connection despite indoor confinement and the disappearance of pigs from the landscape.

The media image of the factory farm that is being cultivated in the above projects, fueled in part by midnight undercover break-ins that take place after working hours, is one of systemic neglect underpinned by a lack of face-to-face attention by workers. Confinement is framed as a labor-reduction process of locking thousands of animals in an unattended barn with automated feeders and letting them grow, suffer, and occasionally die unnoticed (see Imhoff 2010). Exposés such as the HSUS video interrogate industrialization at its apparent core. They suggest that efforts to develop automation technologies and machines to reduce labor costs have resulted in a form of systematic abuse. This chapter, however, develops a distinct reading of modern hog boredom. While it is true that corporations pay less in labor hours per pound of pork than they did a generation ago, the industrial pig, whose body is rife with the side effects of tedium, is not a straightforward product of neglect. For one flipside of industrial automation is the *intensification* of work for the few who nonetheless remain within these sites. Indeed, I will argue that there is a lot of human *and* porcine work underlying these scenes of apparent boredom. The lethargic pig depicted in those videos is not just a product of mechanization: it is a being paradoxically created through and requiring constant attention, of intensely intimate forms of work. Monotony is being actively made. In turn, this chapter's aim is not to lament society's loss of face-to-face attention across species—and, as such, privilege work as an ethic—but instead to open questions about how work is being both reduced and intensified.

In the 1970s, Harry Braverman, in *Labor and Monopoly Capital: The Degradation of Work in the Twentieth Century*, marshaled a classic critique of industrial labor's monotony. His notion of the degradation of work is remembered as a Marxist reading of capitalist management's efforts to plan, divide, and simplify the labor process and render humans interchangeable for the purposes of more easily extracting value from their labor. Industrial work— the assembly line being the iconic example—was built around removing embodied skill from the process, often using machines, to cheapen the cost of labor and decrease people's control over their livelihoods. And the book most certainly is these things, depicting an industrial world where the only means of survival for the majority is through standardized and boring forms of work—one where people are means, rather than ends. But Braverman was also more imaginative than this: his book was a passionate attack on "how the worker, systematically robbed of a craft heritage, is given little to nothing to take its place" (1998, 5). Braverman understood crafts as tying skilled workers to a history of human engagement with the material world, an ongoing dialogue with the past through which a tradesperson's every action implicitly indexes thousands of years of knowledge. Scientific management's efforts to separate conception from execution of work signaled capital's monopolization of annals of human creativity. It was a reduction of humans to a figure of the "animal"—which, for Braverman, meant beings who instinctually do the same thing.

Braverman's book remains problematic. His image of the universal crafts-man excludes most of humanity. The figure of instinctual animality that under-lies his work-based human exceptionalist ideals is reductive. But his notion of monotonous degradation—of simplifying tasks into the tedious repetition of one thing, or deskilling—does help me think about what it means to industrial-ize (and deindustrialize) a hog. For we might flip Braverman on his head and say that it takes a lot of monotonous human work to make actual flesh-and-blood hogs into his figure of the instinctual animal: beings who repetitively do only one thing. Put differently, the industrial pigs discussed across this chapter are doubly overworked. They are complex creatures whose lives are engineered to make them execute one task, such as digest or gestate, in ways that require their physiologies to course ever-more-intensely with intimate forms of human labor. As such, this chapter is less an attempt to theorize how domesticated animals naturally work (see Porcher 2015) as it is an examination of how they are made into work (and very specific kinds of industrial workers). Industrial hog biologies have come to lie "within" human workers' actions as much as they reside in pigs' bodies, just as hogs now attune to and embody industrial human farmworkers in new ways. This tension underpins forms of intimate exploitation and industrial care that everyone who works within factory farms must navigate.

Single-Trait Life and Labor

The events of this chapter unfold in a hundred-mile-radius rural region of the US Great Plains and Midwest, one where sublime numbers of pigs—as many as seven million—are annually born, raised, and killed. Such a capital-intensive intervention into animal life and death has drawn four thousand migrant workers and managers, leaving in its wake a rural region where twenty-six lan-guages are spoken in primary schools. In the early 1990s, a handful of corpo-rations entered this region with the goal of trying to "fully" vertically integrate hogs. They merged historically distinct sites such as genetics, farms, feed mills, slaughterhouses, and value-added post-kill processing while operating through the wage labor of people who held little background in agriculture. But this was not just a matter of farmer dispossession and monopoly; it was not simply a process of taking over every node of the pork chain to extract discrete profits

Figure 3.1. Finishing barn. Photograph courtesy of Sean Sprague.

from each one. Vertical integration was instead about remaking the porcine species into a different kind of biological and capitalist creature. Put differently, these companies' existence is only half premised on making millions of animals. Vertical integration is equally a project of standardizing life: increasing the animals' value by making pigs' bodies more uniform to reduce the labor costs of automated disassembly lines, receive higher prices on pig parts from processors and wholesalers, and make the system potentially scalable and moveable across borders.

It was the execution of this photograph that first forced me to rethink the human-hog relationships that underlie standardized life (figure 3.1). This image is a series of one thousand different frames that have been digitally stitched together. The execution of this shot required the human subject of the photo—a forty-something employee who spends his days driving around the countryside, inspecting animals in different growing barns for new sicknesses—to stand in one place for ninety minutes. When my photographic collaborator, Sean Sprague, needed to adjust the lighting in the barns, he would give the man a short break while I acted as his body double. I was surprised to find that I could not last more than two minutes. The hogs would circle me, nip at my knees, and knock me over. I did not have the bodily habitus to interact with these two-hundred-pound hogs that were four and a half months old and collectively penned; I could not hold my hands in such a way as to corral the animals, stop their nipping, or make them circle rather than charge me. Conversely, one might also say that these hogs could not figure out how to interact with me; they had not encountered a human who held his body in this manner. This was jarring because, after working in breeding farms, I thought I knew hogs pretty well. It dawned on me that I had reasonable knowledge of pregnant sows and baby piglets as they live in individual crates—I could enact forms of labor that allowed us to get on together—but knew little about how to behave with open-penned, near-grown meat hogs. Meanwhile, there are many workers who have likely never encountered a boar, sow, piglet, slaughtered carcass, or rendered fat. Some have gained rich expertise and bodily habitus with open-penned pigs aged four months. But due in part to biosecurity concerns that pig diseases could transfer across workers' bodies (see Blanchette 2015), they labored solely on this stage of porcine life.

These moments open new ways to think about the concurrent division of labor and animality on factory farms. They suggest how there are different kinds of working pig and working human that exist within "the" industrial meat hog that consumers encounter on dinner plates. The project of making a uniform and standardized meat pig for human consumption has paradoxically led to an internal fragmenting of the species across its life/death cycle. There are "genetic sows" with low rates of ovulation that make hardy "commercial sows" who, in turn, generate massive litters of fragile "meat hogs" that are killed in slaughterhouses. There are certain kinds of boars who only provide their semen, and other breeds of Meishan boars refined only for the potent pheromones in their saliva to stimulate sows during artificial insemination. All told, there are some thirteen different breeds, types, and ages of pigs that underlie the meat hog that goes to slaughter. Each of these pigs is selected for one or two traits and provide small parts of themselves to the project of industrial meat; each is monotonously made to work on one small aspect of the American meat hog. Meanwhile, each human worker is specialized into monotonous labor with one kind of hog, or even one aspect of porcine physiology.

The ongoing work of maintaining monoculture in the American factory farm, following Sarah Besky (this volume), is thus a matter of opening up many distinct monocultures *within* the porcine species' physiology. A slaughterhouse worker might make the same slice of the upper left ham 9,500 times every day, knowing the possible distributions and contours of tendons and fat in that one anatomical muscle with a sensory depth that surpasses any textbook. Those who stimulate sows in artificial insemination gain a more profound tactile knowledge of hog reproductive instincts, perception, and sentience than most any animal ethologist. A friend of mine named Robin was a "Day One." Day Ones, as the name implies, work with piglets during the first twenty-four hours after their birth. She heals piglets with unique ailments that block their ability to get adequate sustenance; Robin identifies the weakest animals in the litter—often the runts—and constructs things like duct tape body casts to help their muscle fibers grow. She was constantly seeking new ways of seeing trauma and building techniques of caring for damaged one-day-old pigs. Each effort to heal a unique piglet injury would help her better, as she put it, "save" future generations of hogs born in these barns. Her engaged ethic of care and learned taxonomies of trauma, combined with the sheer number of piglet bodies she had seen in such a space of proliferation, meant that her knowledge of this precise stage of porcine life likely exceeded that of any animal scientist. Workers like Robin are exploited through industrialized hog biologies, but they have also acquired unprecedented expertise with single facets of pigs.

Sentience, or the Sow That Cannot Be Pet

In a conversation at a party with a former farmworker named Juan Marquez (pers. comm., September 28, 2010):

Juan: So . . . what do ya know about pigs?
Alex: . . .
Juan: They have almost three hundred and sixty degrees vision. [He moved his pointing fingers from his eyes to the back of his head.] They are always watching you. Sometimes they look like they are not looking at you. They turn their heads like this [he tucked his chin tightly into his chest]. But if you look at their eyes, you will see they are always following you.

Juan's statement still sticks out in my mind, many years later, and not only because every time I enter a confinement farm, it makes me feel thousands of hogs' sharp gazes following my body. Rather, it was because of an incident that occurred when I was at work shortly after that party.[3] Our primary job in the breeding barns was to artificially inseminate sows. In practice, this means sitting on sows' backs, facing their backsides, and rubbing their sides in ways that try to imitate the mating behavior of boars at the moment of mounting. The goal is to activate a sow's muscle contractions so that she draws in the bag of boar semen through a foam-tipped, straw-like spirette that we inserted into her uterus. On this day, someone miscounted the spirettes we needed. One of my co-workers asked me, the slowest at stimulating sows, to walk back to the barn's workshop to get more so we could finish the day's inseminations before lunch.

As I was walking along the solid concrete path that juts between rows of metal gestation crates, I suddenly noticed a sow that was staring at me. Almost comically so. She was standing erect, and her neck was craned out the top of her cage. Her eyes were fixated on my movements. Without thinking, I paused and reached over to stroke the top of her head. In retrospect, this was likely the first time that I had made skin-to-skin contact with a sow that was not in estrous. Despite being surrounded by over two thousand industrial hogs for nine hours a day, we rarely touched sows except when their bodies were locked in place by lordosis reflexes that cause in-heat sows' leg muscles to freeze once strong pressure is placed on their backs.

"Alex, STOP!" I turned around to see my co-worker Maria, standing twenty feet away. She looked angry, her fists clenched in front of her. Maria was usually the most relaxed of my co-workers on Sow #6. She was known to flout arbitrary rules and managerial authority. Maria was a native K'iche' speaker who came to the Midwest from Guatemala three years previous, and we often shared conversations in Spanish, but this was the first time I heard her speak English. "Don't touch the sows!" she screamed. My violation of the unspoken terms of industrial animal husbandry might seem innocuous to most readers. I pet a sow; I tried to interact with a hog in the same way that I unthinkingly approach all animals based on owning a domesticated dog when I was a child. Maria relaxed as she saw the shock register on my face. She gently explained that petting sows could surprise them. Such a gesture might cause the sow to startle or writhe in her cage. The sow might get upset and bellow, alarming the rest of the herd. Such "excess" affect, Maria's words suggested, a deviation from monotonous existence, could cause the sows to abort, potentially resulting in a wave of miscarriages as ripples of excitation pass down a row of animals in gestation crates. "[These sows] are not very strong," she said.

There is a lot to say about this fleeting incident. In spite of its ephemerality, it hints at a striking picture of how confined animals may be evolving. Granted, in all my time in factory farms—whether as an entry-level laborer or shadowing managers in their administrative tasks—I have never encountered a rule that says you cannot touch the hogs. Indeed, I don't even think the single act of petting a sow was such a huge deal in and of itself. Rather, I think that Maria was using the incident as an instructive lesson, to let me know that my actions can have repercussions for animals' well-being. Still, as a casual stroke of the back potentially threatens reproduction, one gains a glimpse at the radical corporeal weakness of industrial animals. If we follow this workplace sensibility to its conclusion, then any deviation from the norm—any excitation, any sudden change in the porcine sensorium—has the potential to cause a miscarriage among these imbalanced animals that have been genetically selected primarily for their reproductive potency. The emergence of a sow threatened by forms of novelty is perverse precisely because hogs that live outside of indoor confinement tend to be attracted to novel experiences (Grandin 2005). Yet in this highly regimented space, nonuniform events can apparently threaten sows' capacity to proliferate flesh. This is a sow that is engineered to be frail, an embodied vulnerability that paradoxically results in the porcine species becoming more finely enmeshed with human action.

We are left with a sow that cannot be pet—or, one that would have to be pet the exact same way, at the same time, every single day. This is an animal whose very welfare is predicated on a determined lack of individualized attention. In this workplace orientation, taken seriously by even blasé workers, the quandary is erratic humans whose every laboring intervention is a potential source of singularity for the porcine sensorium. This is not a matter of treating pigs as mindless widgets, as one might imagine the factory farm is wont to do. This factory is about organizing human and other-than-human affect. It is not a place devoid of feeling, but where feelings become enfleshed. It is a workplace orientation whereby one tries to experience one's self from the sow's view, as one's body emits varied signs that can unintentionally affect hogs.

If, as Eduardo Kohn (2013) has argued from the vantage point of rainforests teeming with interspecies interaction, to be alive is to exist embedded within and to produce meaningful signs, to be unfolding in encompassing webs of signification, then this horizon of the factory farm is striking. It is a proposition of industrializing the semiotic interfaces that extend between and within species, of the signs that pass across and stitch together beings. These sows suggest a factory farm that is not genetically reductive but one where all signs that compose pigs' routine lives in the everyday—auditory, visual, or biological—come to be mediated by labor. That is, and pushing beyond the use of uniform diets or genetics to create predictably uniform meat, this vision of a factory farm is premised on the production of monotony through labor that regulates porcine emotion and the hog's perceptual world. The ideal animals are those that experience no novel events even though they have become hyperattuned to, and even biologically absorbing, the physical work of human others. Or, put differently, this hog is a being that could turn any minor human behavior potentially into an act of work in the sense of an action that affects a product.

My sense, however, is that this hyper-attunement of sows to human behavior is also inseparable from the state of current porcine genetics. The past fifteen years have seen drastic changes in how the porcine species manifests within the factory farm, marked by a qualitative change in animal biophysical nature through its quantitative overproduction. As genetic "improvement" has fixated primarily on ovulation rates, litter sizes have been increasing—from an average of ten pigs per litter in 1990 to over fifteen per litter on the most "advanced" farms when I was conducting research—such that they are outpacing the sow's capacity to supply nutrients in utero (see Miller 2007; Blanchette 2019). The side effects of this vital declension can be multiple: a meat pig chronically lacking in nutrients; increased rates of stillbirths during unsupervised farrowing; larger quantities of runted infant animals; bony, lean sows rendered even more prone to injuries such as bedsores due to added lactation demands; and sows that easily abort. As some farms have moved to implement twenty-four-hour shift work to guarantee monitoring of fragile life forms during moments of delivery, bio-industrialization of ovulation is paradoxically requiring more—or, at least, more constant and consistent—labor time. In the process, what emerges is the suggestion of a species whose very life requires standardized human labor. The sow cannot be individually pet, but it must be uniformly worked.

This hog's curious state of embodiment thus demands that we pose the question of what human labor is becoming in these spaces. For this is not a matter of labor as human creativity, of making things, but instead one that merely maintains another organism's capacities to gestate (see Gorz 1994, 56; Besky, this volume). Conversely, what kind of behavior is not labor once every human action is potentially a matter of performing for industrial pigs—of generating and doing porcine sentience within one's actions? Minimally, this little incident suggests how engineering breeding sows to primarily do "work" on one thing—in this case, laying on the ground undisturbed while gestating increasingly larger litters—makes other forms of their existence, such as perception or the eroding epidermis, into terrains of now-necessary human labor.

Hormones, or Lutalyse

As part of Sow #6's training program, I learned the tasks on the delivery side alongside Raul, a wiry and wisecracking man of some fifty years of age. This was Raul's second job since he had undergone a self-imposed exile a year previous. He was trying to keep the mortgage on his home afloat after he lost his job during the collapse of Miami's construction industry amid the 2008 US recession. Grensome Meats, a nearby beef plant, flew Raul out to the area. They put him in temporary housing, gave him a $500 advance, and required him to work on the cut floor for at least one year. The second he could quit, Raul moved south with some friends. "My body's too old for those slaughterhouses," he explained, as he flexed his right hand at the memory of pain in tiny muscles from repetitive motion with a knife. He did not speak the requisite English to be employed in retail or in higher-paying oil fields. "The only work they'd give me was in sow farms," he said with a shrug. Sow #6 is not an easy place to work. During my time there, four people arrived for one day of work only to never show up again. Raul and some of my other co-workers figured they were disgusted by the thousands of hogs' concentrated feces. But as I got to know Raul, I realized he derived an ironic pleasure from being surrounded by pigs in spite of the overwhelming stench. He once said he had saved money in Cuba to buy a cow, and that it was a source of great pride to him. "One cow, one cow!" he chanted a few times, cackling at his former sense of value, as he dripped milk into a runt piglet's mouth surrounded by hundreds of animals.

Nursing weak runt piglets is a grating task reserved for new hires. For a week or two, until someone else was hired, Raul and I moved through the farrowing rooms, taking turns to refill and microwave bottles of powdered milk mixed with tap water, coming back to clammy piglets every fifteen minutes with more liquid in an effort to keep them alive. What I did not realize then was that Raul and I were among the first generation of workers to be assigned as human wet nurses. While powdered milk has been kept on confinement farms for decades, for the rare situation in which a piglet was too weak to get sustenance, it is only since the late 2000s that "runt feeder" became a position (see Mavromichalis 2011). Tied to how litter sizes have grown larger than can be nursed by sows, feeding runts reflects how the biology of the industrial pig is increasingly noncontiguous with its body.

One morning, as we were making our feeding rounds, the squeals of piglets were drowned out by a loud BANG! that sounded like an aluminum baseball bat slamming against concrete. Raul located the source of the noise first: a four-hundred-pound pregnant sow at the far end of the room was repetitively swinging her skull against one of the metal bars of her farrowing crate. BANG! BANG! BANG! "What's she want?" asked Raul as we walked around the crate, trying to keep our distance from this seemingly crazed animal. BANG! BANG! BANG! We positioned ourselves on the sow's right side, following the leftward direction of her smashing head to see if the trajectory of her motion pointed us toward something. Thinking she was hungry, I ran to get a scoop of feed. The sow did not acknowledge me. Raul checked her water nozzle. It worked fine. We sprinted down the hall outside of the farrowing room, our co-workers giving us puzzled looks as we passed by them. Finding our manager taking inventory in the workshop, Raul shouted, "There's a crazy sow!" Without thinking, I blurted out, "Please go check on it! I think it's trying to kill itself!" After returning from checking on the sow, my manager almost sheepishly explained, "This is what happens when we give the Lutalyse. The sows try to nest, but they don't have that here [straw]. They can't nest."

To frame the industrial animal as constituted by standardized (and standardizing) labor might be counterintuitive given that we rarely hear of the diverse hands that underlie meat. Many instead describe how the indoor life of the American hog is riskily sustained with antibiotics mixed into feed, which keep hogs growing despite cramped conditions (Wallace and Koch 2012). The pig is fed a cocktail of lincomycin, zinc bacitracin, tetracycline, sulfamethazine, and penicillin matched with scores of vaccines, vitamins, or beta-agonists that promote muscle leanness. Such drugs maintain the industrial species in a dual sense: sustaining its raw life amid barns teeming with illness and regulating its carnal composition with increased and semi-predictable rates of growth and fat-to-muscle ratios. But little public discussion has been paid to another ubiquitous class of chemicals: the drugs for labor. These injections are constant on factory farms. They synchronize hormones across groups of animals. Enacting a chemical massification of hogs, hormonal drugs compress biological time into industrial time while temporarily manifesting factory farms' ideal state of animality: a herd of identical sows.

As a factory farm worker, I rarely encountered antibiotics except during a feed mill visit, where hundred-pound bags ready for pelletizing are stacked on pallets. Occasionally, we would inject vials of milky-white liquid penicillin into the necks of acutely sick breeding animals. But many mornings instead were consumed by carefully working on animal hormones with plastic syringes, disposable needle tips, and thirty-milliliter vials of liquid drugs in hand. We would slowly, slowly move behind animals in their gestation crates to avoid provoking any sudden reaction. We would cover the glint of the needle, which can scare the animals, with our latex gloves. Alarming a sow at this crucial moment can result in her forever reacting with rage when a human worker enters her gestation crate, her memories rendering her permanently untreatable. Animal welfare researchers have also found that sows who are scared of human workers will produce smaller and less robust litters (Hemsworth 2003).

Some drugs ensure, for example, that sets of sows will enter into estrous on the day that they are slated for insemination. Others, such as injections of oxytocin, intensify sows' uterine contractions so they deliver piglets in the supervised nine-hour working day. Exaggerating mammalian processes into tight temporal windows, these laborsaving and intensifying drugs end up acting on the injectors to compel workplace action. They reduce the amount of wage-labor time that corporations must pay, while intensifying the pace of work for the few laborers who remain in these spaces. Injections exposed us to animals that can be worked on in industrial time, while ensuring that most porcine behaviors—including right down to the stimulation of hormones— were partially mediated by human action. Indeed, the very idea that there is one person who serves as the hormone worker every morning is an illustration of how formerly autonomous porcine qualities, which once "belonged" to and were generated by the porcine species, are becoming alienated from pigs' vital behaviors and turned into terrains of alienated human labor. As an example of how the pig is overworked, it reflects how new dimensions of hog biology are constantly being made subject to regimented work—in ways that also require human working subjects to shoulder yet more aspects of pig biology.

Of the many drugs we injected into hogs, the vials of Pfizer's Lutalyse (aka dinoprost tromethamine, a prostaglandin) were the only ones that elicited strong reactions from my co-workers. Lutalyse is a birth induction drug that ensures sows will farrow no later than 114 days from the moment of conception, the injection timed so that birth ideally takes place during the eight-to-five working day. Mortality rates are reduced when workers are present to pull piglets out of the birth canal every fifteen minutes during farrowing—a task called "sleeving," reserved for Day Ones such as Robin. Lutalyse was both cherished and loathed on farms. Pregnant women, such as Maria, were forbidden from entering the drug storage closet. Senior workers refused to touch the drug, claiming that contact with the vials made their throats constrict. On the other hand, it was also the most "moral" of drugs. The managers of Sow #6, deemed the most community-centered of factory farms in the area, seemed to use Lutalyse at a higher rate than did others. Managers were proud that they could ensure their employees would have a regular eight-to-five job Monday through Friday, while other corporations' work regimens were structured around the temporal unpredictability of sows' contractions and extended into evenings. Many co-workers left other companies for this one as they were starting families, wanting to know in advance that they could pick up their children from daycare and participate in the kinship rhythms of standardized industrial time (see also Parreñas, this volume; Satsuka, this volume). Whether through bearing similar mammalian bodies or competing temporal interests, Lutalyse was distinguished for creating types of farmwork at the same moment that it manifested hormonal states in animals; the induction drug materialized moments of cross-species proximity and competition. Seaming livelihood and ways of life across species, Lutalyse at once intensified labor in the eight-to-five working window, while also providing an escape from labor on an animal that is now beginning to require almost constant supervision. A means of corporations coping with the growing working demands of the industrial pig—while not having to pay more *waged* labor—Lutalyse could be interpreted as a precarious tool to maintain industrial time amid a hyperindustrial animal biology that now exceeds it.

Meanwhile, the sow smashing her head against the metal bars of a crate was a rare expression of stereotypic behavior, defined by many ethologists as repetitive actions that are invariable in form and serve no obvious function in terms of play, work, or bodily reproduction. Left to their own devices outside of confinement, sows will make a nest out of foliage before they settle down to give birth. Locked in a metal and concrete farrowing crate without straw, their nesting instinct—violently exaggerated through injections of prostaglandin—can occasionally transform into rare behaviors such as head weaving that happens to collide against crates. Applied ethologists tend to explain the stereotypy as occurring either because of an environmental blockade to evolutionarily ingrained behavior, or as the behavioral product of an environment without any stimulation. The animal welfare scientist David Fraser (2008, 142–44) complicates these standard definitions of the stereotypy. He suggests that scholars might overlook the function of a stereotypy. Rather than being useless noise, some have found stereotypies can reduce an animal's level of emotional arousal, calming pigs in stressful situations. But regardless of whether it is grasped as a matter of being frustrated, bored, or as a defense mechanism, the stereotypy can also be read as a manifestation of a being whose evolutionary behavior is estranged as labor. Stereotypies are expressions of being overworked.

Conclusions

In recent years, anthropologists and geographers have initiated a dialogue on the relation between nonhuman nature and capitalist work (e.g., J. Moore 2015; Tsing 2015; Beldo 2017). Resisting the chauvinism of prior labor theories that posit human work as the prime mover of the world, as the singular source of worldly *poeisis* (e.g., Vogel 1988), these authors argue that capitalist value depends on corralling unpaid actions of nonhuman beings (not to mention the unpaid work of many human beings). That pigs give capital "free gifts"—or, that capital shapes nonhumans to appropriate their unrecognized work (J. Moore 2015)—is a compelling way to think about factory farms. Pheromones, gestation, digestion, or skin cell regeneration are all biological processes that are not "authored" or "made" by capital or wage labor, even if they are necessary for the latter to accrue value. But the factory farm is also constituted—as the source of its standardization—by an attempt to reject these "gifts." Agribusiness's ideal, however impossible this may be to fully realize, is one where every biological function of the porcine species is tinged, adjusted, and modified through human work. We might say that agribusiness has its own theory of porcine labor: all pigs work, but they do not work the same way. Standardized porcine life is embodied and enacted by human workers as much as it is by pigs, and the process is still ongoing as more and more dimensions of hogs become work.

   This chapter has been written in solidarity with those critiquing the productivist tedium of confined animal lives. Where I diverge from popular readings of lethargy is when it is premised on the idea of a mechanized lack of work— or little interspecies contact, more generally—as the basis of animal suffering. Instead, the American hog is "bored" through monotonous and intense overwork. The factory farm is a space where porcine states, vital functions, and self-sustenance are being taken over by some, and forced onto others, as routinized labor to make pigs uniform at large scales. Or, at least, that is one half of this situation: one where migrant workers are embedded as life supports into porcine physiologies. In turn, this allows the "specialization" of distinct hogs into breathers, metabolizers, or gestators. Put differently, what it means to be a human worker on factory farms is inseparable from *how* pigs are put to work.

Though it is tempting to think of new dimensions of pigness becoming terrains of labor as a story of increasing capitalist control, it is perhaps better to say this: factory farms are totalizing because of their very fragility. Efforts to make standardized confinement work (out) seem to require unendingly new depths of care and expertise, from skin work to the ethical love of Day Ones. What this points to is a rather different set of questions. How might we write an industrial agriculture that makes political the historically unique expertise and exploitation of farm workers—rather than dwelling in their bodily or mental suffering, or lamenting their alleged deskilling and society's attendant loss of agrarian knowledge? What does it matter that the factory farm—a site of profound exploitation—also generates radical forms of interspecies knowledge and modes of attention, expands sensibilities of what it is to relate to nonhuman animals, and creates remarkable modes of human-hog interdependence? The question is what farm labor politics would look like if visions for alternative agricultures started not as repudiations of exploitation but instead at the emerging commons of expertise among workers and the concentration of historically unique forms of animal knowledge in the hands of people like Luis, Robin, and Maria.

From another angle, however, what remains troubling is the mundane but still remarkable fact that it has become challenging to encounter, know, and form relations with many species outside of work. The modern pig is overworked not only because its genetics ripple with deep histories of labor—in ways that seem to necessitate unending futures of yet more human labor to maintain its basic physiologies—but also in the sense that social knowledge of every organ, chemical function, or trait of this organism has been acquired primarily within capitalist workplaces. The task is thus perhaps not only to put pigs and people to labor in new (or older) conditions—for instance, on open fields and pastures deemed more natural (see Weiss 2016). Instead, and following Weeks (2011, 151–74), the events of this chapter make me think about the need to rekindle the general spirit of the eight-hour day for changed ecological times and relations: honing efforts to work against work, actively struggling to reduce work, and creatively privileging other dimensions of what it means to be alive. Yet what the industrial pig suggests is that this is no longer a project that can be exclusive to human workers alone. As deep histories have shaped organisms and ecologies to work, or otherwise become dependent on human labor, imagining a life not dictated by the demands of labor—even just for humans—appears to be a matter of relearning and shaping how we relate to nonhuman others anew.

Notes

1. Company and person names in this chapter are pseudonyms, designed to provide a degree of anonymity for the many executives, managers, and workers with whom I worked. Similarly, I am unable to state with precision the region of the Great Plains and Midwest where this is taking place. Each major pork corporation centers its operations out of a distinct state. Similar operations to these ones exist in Illinois, Iowa, Kansas, Missouri, Nebraska, Oklahoma, Texas, and Utah, in the United States, and Manitoba, Canada.

2. Industry studies usually treat bedsores as a technical problem of the design of both barns and pigs (e.g., Rioja-Lang, Seddon, and Brown 2018). The lack of straw bedding, the pointy corners of concrete slats, and, most of all, the leanness of genetically selected breeding sows are certainly contributing factors to the ways that bones and flooring pinch against the epidermis. But I will also be arguing, across these pages, that they are symptoms and manifestations of interspecies labor.

3. See Blanchette 2018 for a very different description and analysis of this incident, one that explores what it might teach us about the meaning of "animal welfare" for confined hogs.

Part Two

Labor Struggles

CHAPTER FOUR

The Job of Finding Food Is a Joke

Orangutan Rehabilitation, Work,
Subsistence, and Social Relations

JUNO SALAZAR PARREÑAS

Making a living in Sarawak on the island Borneo, as it is in much of the Malay-speaking world, is to *cari makan*. It literally translates as "to find food."[1] The words together suggest a shift from subsistence to wage labor, one that hints at the ambiguous distinction between agricultural cultivation and what some would call hunting and gathering, or "food collecting" (Ellen 1999). Both words in Malay, "cari" (to search, previously spelled in Romanized Malay as *chari* or *charek*) and "makan" (to eat), are generative and blend the literal and metaphorical. In 1887, British colonial administrator Frank Swettenham (1887) offered a definition of "makan" as eating, devouring, dining, consuming, and penetrating, such that one could *makan gaji* (to work for hire; literally, to eat wages), *makan rachun* (to take poison), *makan rumput* (to graze; literally, to eat grass), and *makan suap* (to take a bribe; literally, to eat a mouthful of food). "Cari" by 1908 meant seeking for, looking for, and searching for anything, including a "prostitute" (*perempuan menchari*; literally, to find a woman) or a source of livelihood (*pencharian*) (Wilkinson 1908, 35). The particular idiom "cari makan" might have especially circulated among Peranakan port cities of the Straits Settlements in the twentieth century or perhaps earlier: Penang, Melaka, and Singapore—the latter being the largest entrepôt near Sarawak across the South China Sea. In a Baba Malay dictionary penned by retired pharmacist and independent historian William Gwee Thian Hock (2006), the term "charek (cari) makan" is a reference for earning a living along with *charek kaki* (to look for one's peers; literally, to look for feet) and *charek kreja* (to look for a job). Today, throughout the federal state of Malaysia, founded in 1963, and in Singapore, which separated from Malaysia in 1965, "cari makan" is a common phrase for discussing an unsatisfying but tolerable means of employment. For instance, if you ask a friend if they are enjoying their new job, they might simply reply, "Cari makan, lah" (Lee, n.d.).

Finding food plays with the literal and metaphorical dimensions of subsistence, which was a crucial keyword among an earlier generation of anthropologists. What might it suggest, then, when the idiom of "cari makan" emerges as a new expectation placed on Sarawak's displaced orangutans? Orangutans that end up in Sarawak's wildlife centers, like the people employed to care for them, are jokingly expected to cari makan, or earn a living. But like all statements that can be dismissed as mere jokes, they are deadly serious (Caton 2005).

Once at an orangutan conservation conference hosted by the state of Sarawak, the welcoming address included a joke about the idiom "cari makan." The laughter it received from audience members suggested that some of the crowd found it funny. Or perhaps the enthusiastic laughter was a sign of respect for the jokester. The joke was delivered by a VIP *datuk* (an honorific title) who had two important appointments. Firstly, he was the deputy permanent secretary of the Ministry of Planning and Resource Management, which was the state agency responsible for some of the most controversial development initiatives in the history of the country. Secondly, he was CEO of the Forestry Corporation, which was the semi-governmental agency that was formed when the Department of Forestry was partially privatized by the state legislature in the late 1990s. The Forestry Corporation manages the state's parks and biodiversity conservation and issues timber licenses.

The datuk's joke was this: a university student wore an orangutan costume at a zoo for a job because he had to cari makan. One day, he fell into the tiger pit. At the moment it looked like the tiger was going to eat him, he realized that the tiger was another student, who was also in costume. The student in the tiger suit said to the other student in the orangutan suit, "Sorry, pal, but we all have to find food."

In this joke that fits the American idiom of a "dog-eat-dog world," where finding a livelihood means competing with one's peers to the point of outright killing a friend to make a living, the datuk found a connection to the conference's theme: meeting orangutan *and* human needs. The joke made the point for the datuk that orangutans' needs and humans' needs are in competition with one another. It went without question who comprised human here, what differences might be glossed over with the singular term for human, and how humans of Sarawak may very well have intensely different perspectives from one another. It also went without question what exactly constituted a need and how that might be different from desire or perhaps even greed—a greed for land, timber, sand, and other natural resources that could be exploited where wild orangutans and upland Sarawakians dwelled.

Aside from such unspoken critical questions that lingered in the air-conditioned conference room of a newly built world-class hotel, the concept of cari makan is a crucial one for understanding orangutan rehabilitation today. Orangutan rehabilitation sites like Lundu Wildlife Center, its older sister Batu Wildlife Center, and their predecessor site from the colonial era managed by the Sarawak Museum are places to cari makan. They are places to figuratively and literally find food for displaced orangutans and their displaced caretakers; where Ching, the semi-wild adult orangutan and other displaced orangutans like her who had been taught by humans to survive, can sustain herself by finding raw sweet potatoes specifically left for her on an ironwood platform or wild figs growing in the forest during fruiting season; and where the orangutans' caretakers could perform their job and earn a wage for it. For people who are paid to rehabilitate orangutans, the paid work competes with the labor of growing and harvesting rice, similar to the way that the first orangutan rehabilitation workers in the 1960s were put into a position of having to choose their paying job over their personal desires and social obligations.

The hypothetical zoo in the datuk's joke, the two working wildlife centers in Sarawak, and their predecessor, the orangutan rehabilitation project run by the Sarawak Museum in the 1950s and 1960s, are all "workscapes," borrowing the term from environmental historian Thomas G. Andrews (2008). The wildlife center and the rehabilitation center give a different feel to the question of how nature works by getting us to think about the fuzzy boundaries between wildness and cultivation through the pressing question of how to make a living under new circumstances, for both people and the nonhuman primates under their care. In this respect, it invites revisiting what subsistence might mean in the twenty-first century with human-animal relations in mind.

In this chapter, I show that orangutan rehabilitation is a story of how semi-wild orangutans and people caring for them must earn a living under conditions of wage labor and the workday. In short, it is a story of how nature is made to work and how workday hours transform social relationships. Wage labor here serves as a synonym for holding a job, one that is compensated according to temporal increments like work hours or potentially a salary. "Cari makan" in particular emphasizes the relationship between labor and food. Orangutan rehabilitation—from the time of its experimental beginnings in midcentury Sarawak to the present as a private-public partnership staffed by subcontracted workers—imposes regimentation of both time and sustenance. This phenomenon is reminiscent of the anthropological study of shifts from agrarianism to developed economies (Comaroff 1985; Ong 1987; Taussig 1980). But what makes this situation striking is the way in which nonhuman animals are induced into the same regimentation in terms of hours and diet.

Anthropological scholarship has recently highlighted the role of compensation in a world without work. For instance, James Ferguson (2015) illustrates the South African approach to mass unemployment, which has been to compensate the nation's most vulnerable citizens, particularly senior citizens and those rearing young children. He argues for expanding compensation to young men, a demographic increasingly unemployed. Tania Murray Li (2014) highlights a similar dilemma in Indonesia in which the exploitation of land for cacao has meant a transformation of social relations where family and village relations are jeopardized in the face of inequality. While both consider the limits of capitalist assumptions of continued growth and expansion, the stories contained in this chapter suggest not a contraction but a surprising expansion of the logics of labor and earnings. Here, it is not a world without work but a world where everybody works.

A world in which everyone works seems to be a realization of a fantasy shared between functionalist anthropologists of yesteryear who were concerned with social reproduction and continuity across generations and laypeople influenced by historically Western European ideas of nature as a mechanistic model of harmony, exemplified by bee colonies, where parts work together in unified and complementary divisions of labor (Segal and Yanagisako 2005; Merchant 1980; Kosek 2010). However, the world of orangutan rehabilitation is far from a materialization of harmony in which nature simply works. Rather, tensions disrupt workplace relations—for both people and orangutans who convey discordance with being put to work. Nature does not simply work without some grievance about what the work conditions are and how exactly work is to be compensated.

For a generation of anthropologists well into the 1980s and 1990s, subsistence was the answer to questions of how to live, how to interact with the environment, and how to reproduce social relations. As June Nash simply phrased it in her comparison of Maya villagers in Chiapas, Bolivian miners, and laid-off industrial workers in the United States, subsistence is "the welfare problems faced by women trying to feed their families and keep their children alive"

(1994, 9). Subsistence is about survival and gaining enough to eat, regardless of where somebody lives and regardless of whether one forages, grows, or buys food, or any combination thereof. For Renato Rosaldo, writing in the 1980s, the "routine subsistence techniques usually portrayed in 'ethnographic realism'" (1986, 98) were not as interesting as the cultural notions that motivate, inspire, and compel people to keep living. Rosaldo hypothetically asks, "Why [do] the hunters portrayed in most ethnographies bother to get up in the morning and face nothing but routine drudgery?" (102). Roy Ellen (1999) also critiqued the hegemonic understanding of subsistence among many anthropologists and human ecologists of his generation as being too focused on tools to the detriment of recognizing that technology also includes intangible knowledge. Yet he continued to theorize subsistence by offering a more precise definition of subsistence as the "aggregate of extractive processes" (Ellen 1999, 95). Following Kregg Hetherington (this volume), if we recognize that agriculture is about killing, whether harvesting, slaughtering, burning, cutting, or weeding, then we should also recognize that the aggregate of extraction includes extracting energy from deathly agriculture as well as extracting time spent seeking and eking out sustenance.

The extraction of time differs from how primatologists generally conceive of caloric expenditure for caloric intake; namely, the time and energy spent foraging (Knott 1998). The difference lies in the form of sociality: the orangutans in wildlife centers are subject to human social relations. In this regard, it can be understood as metabolism in the sense of its relational exchange of matter, as Sarah Besky and Alex Blanchette explain in the introduction. Time is not quite matter, and yet time is compensated through material food items. This form of human social relation taking shape within Sarawak's wildlife centers differs from the rhythm of the rhapsody described by Shiho Satsuka (this volume) insofar as this is not a story of nonhuman agents seducing humans into nonhuman temporality or arrangement. Rather, this story is akin to the biochemical induction of birth among industrial pigs in the United States described by Alex Blanchette (this volume) that is meant to coincide with the regimented schedule of hyperspecialized human workers. Yet regimentation only goes so far. Like sick tea plantations that suddenly close, as described by Besky (this volume), or heat that cannot be willed away (see Alex Nading, this volume), humans do not fully control rhythms either.

In the world of orangutan rehabilitation and human-orangutan social relations, spontaneity and surprise often thwart attempts at clockwork consistency and regimentation. In a way, the uncertainty and indeterminacy that characterize orangutan rehabilitation mirror the conditions of orangutans undergoing rehabilitation in wildlife centers. These orangutans are, in the words of workers and other practitioners, "semi-wild." This is not a re-wilding, since there is no delusion that the wild can resume nor is there a hope that this area will be re-indigenized with introduced orangutans that are actually native to other parts of Sarawak (Rose 2004; Lorimer 2015). "Semi" is not quite the same as "quasi" or "pseudo": the partiality of being semi-wild entails foraging and eating cultivated food as well as having freedom of movement, interaction, and encounters within constrained conditions. There is an earnest attempt to achieve states of semi-wildness within the physical limitations of the parks situated just beyond the city of Kuching, one of Borneo's largest cities.

This chapter first examines the staffing needs of early orangutan rehabilitation efforts in the 1950s and 1960s, which revealed an arising problem concerning managers' demands for twenty-four-hour surveillance of and companionship with semi-wild orangutans, which conflicted with workers' desires for leisure and human sociality. It then considers how semi-wild orangutans at today's Lundu Wildlife Center became constituted as semi-wild through a regimented time frame composed of nine-hour workdays. It then considers the foods that sustain the labors of both orangutans and their keepers today, the literal food (makan) that they seek (cari). By treating the metaphor of cari makan literally and materially, this chapter dually examines orangutan rehabilitation sites as workplaces and the practices of orangutan rehabilitation as work for both people and orangutans. By recognizing the forms of labor that constitute orangutan rehabilitation as transformative for everybody at the site, including nonhuman subjects, it offers up the question of how labor is compensated when everybody must cari makan.

Making Merry instead of Work

Orangutan rehabilitation as a form of work demanded that its workers develop a new relationship to both time and social relations from the very beginning in the 1950s. The idea of orangutan rehabilitation began in the colonial home of Barbara Harrisson, who had been a volunteer at the Sarawak Museum before she divorced her forester husband and married the museum curator, Tom Harrisson. From the initial experimental efforts of the project of rehabilitating orangutans, we see how efforts to teach displaced orphaned orangutans independence is inextricably tied to a project of disciplining Indigenous labor (Parreñas 2017, 2018). The work of training is seen as a means of instilling independence within the constraints of an encroaching capitalist modernity.

The small grounds of the Harrissons' home, Bungo Segu, with its durian trees, pineapple bushes, hibiscus plants, and a profusion of native cultivars growing on their own with rare grounds-keeping interventions, became the place for orphaned orangutans to go once they were confiscated by the colonial state. Barbara lived there with Tom, their Malay housekeeper, Amah, and Bidai, a Selako youth whose father she described as a "Grand Master" of a shaman (Harrisson 1962, 36). She considered Bidai's father a "distinguished friend," and he had asked the Harrissons to accommodate his son so that he could "see and learn a little in town" (Harrisson 1962, 36). He received a wage from the museum for being the key animal handler in the Harrissons' rehabilitation efforts. Bidai's days involved cleaning the orangutans' sleeping areas and climbing trees with them. His first wages were spent on getting a transistor radio and full trousers: accoutrements of civilized urban life.

It was on the grounds of the Harrissons' home that orangutans like Eve and Bob learned to use their limbs to walk in addition to climb. It was where they lived off a diet of fruit and reconstituted cow's milk thickened with rice. It was ultimately where these orphaned orangutans faced material limits to independence, as constrained as it was within boundaries of the property. The more than thirty orangutans that were eventually hand raised at Bungo Segu, except for the last two, were all sent to zoos. Under the Harrissons' directorship, the receiving zoos were located in such places as the United Kingdom, the Netherlands, and Germany.

It took much effort by the Harrissons to convince the colonial state and the Department of Forestry to help realize a more tenable response to the problem of orphaned orangutans. By then, orangutans were of increasingly global conservation interest. The Department of Forestry capitulated to the Harrissons' efforts and granted land in 1962. That land was located in Bako National Park, the first national park of Sarawak, which is on a mangrove coast but has seven different kinds of ecosystems. The park boundaries spanned twenty-seven square kilometers on a peninsula at the South China Sea, at the deltas of the Kuching and Bako Rivers. However, most of this land was uninhabitable for orangutans. One of the ecosystems, the mixed dipterocarp forest, could in theory support orangutan life and behavior, even though orangutans did not appear to have a history in that area.

The site became operational by June 1962. By August that year, the Sarawak Museum files address the issue of disciplining Sarawak Museum workers at the site. One night in August, Bidai and two of his co-workers snuck away at half past eight to attend a wedding at Kampung Bako, the closest village to Bako National Park, which was across the bay. It required taking the museum's boat across the open waters of the South China Sea.

A handwritten letter written by a fourth co-worker reported to their boss that the three used the museum's Seagull motorboat and that he saw the three "married men." The English idiom of "making merry" remains a contemporary euphemism for drunken pleasure in Sarawak. Even though Kampung Bako is a Malay, and thus Muslim, village, this occurred at a time when festivities, including Hari Raya (Eid-al-fitr), were celebrated with rice wine and rice whisky throughout Sarawak across different ethnic groups and against what has now become entrenched religious prohibition.

On the backside of the handwritten letter, Barbara Harrisson jotted the following list:

- No authority to leave Limau unless emergency.
- No authority to use Seagull except for emergency and during drought (for water fetching)
  irresponsibility:
  leaving the station (stores, animals, etc. . . .)
  taking a small boat to sea with 3 people in
  Explanation?
  Bidai is getting the sack as responsible leader if explanation cannot be given.

Details of this episode, found in the museum archive files, show that Bidai at that moment faced two competing obligations.[2] On one hand, he and his co-workers were personally invited by the groom's father to attend the wedding festivities. On the other hand, he was obliged to work around the clock at a field site. Orangutan rehabilitation involved a lot of work. It not only entailed being present with the orangutans but also managing their food supplies as well as the caretakers' own food and water supplies. It required tearing themselves from human social relations of choice and birth to instead confine themselves to social relations with only co-workers and orangutans.

The next document in the file was a report from Barbara Harrisson. She wrote:

All three admitted having gone to Bako on Saturday night, Aug. 3. Each of them received an invitation from a family at Bako in celebration of the marriage of their son.

The temptation was too great. They left camp at 8.30 p.m. and returned at 3:00 a.m. on Sunday, reaching camp at 5:00 a.m.

They did not give any excuse. They went because they thought they could get away with it.

They admitted they were wrong and asked for forgiveness.

I particularly told Bidai that he is liable to be dismissed for his irresponsibility as he is in charge of the camp. He said that was his first serious mistake since he joined the Museum and asked Curator to be lenient on him

Conclusion:

Their silly act was not just an impulse. They [had] contemplated and talked about it even before Jawawi [the fourth worker and complaint lodger] had left.

No written record documents how Barbara and Tom Harrisson deliberated on the course of action to take. However, Bidai ultimately did not get fired. Was it the years of service to the museum and to the Harrissons with only one misstep that compelled leniency? Or perhaps was it that Bidai at that point was the most experienced person working with orangutans and was thus de facto irreplaceable? Or could it be that his apologetic attitude appeased the Harrissons' demands for discipline? We may never know since both have since passed away, but we do know that the work of orangutan rehabilitation even in its early days had the expectation that one would prioritize social bonds with orangutans over social bonds with humans around the clock and every day.

It could be that the physical limitations of the camp emphasized the isolated conditions of this workscape, and that isolation extended into Bidai's social life. Perhaps Barbara Harrisson felt compelled to make such demands given the newness of this experiment in raising orangutans to become semi-wild. Nothing of its kind had been attempted before. Yet another reason could be simply material insofar as food supplies had to be guarded from proboscis and macaque monkeys. All of these workplace conditions—the isolation, the undivided attention required in maintaining an ongoing experiment, and the job of protecting food stores and other resources—give a sense that this job was perhaps not a source of complete satisfaction for Bidai, who earlier intended to "see and learn a little in town" (Harrisson 1962, 36). Bidai seemed to be learning the lesson on the job that this work was just "cari makan, lah." This was proving to be unsatisfactory but tolerable work. Perhaps that was the lesson of modern, industrious ways of living.

An Orangutan's Work Day

In the late 2000s, Ricky, a waged employee at the sanctuary, and I sat on a feeding platform amid the bananas and sweet potato left out for Ching the orangutan to eat in case she was unable to successfully forage. This was Ching's job, as Ricky joked:

> Their job: sit, sleep, travel around
> Until 5 or 6 o'clock [a.m.]: sleep
> 6 o'clock [a.m.]: travel around
> 7 to 11 [a.m.]: relax
> 11 [a.m.]–1 [p.m.]: play on the ground
> 2 [p.m.]: travel around and eat until it's time to sleep again.

To be semi-wild, Ricky seemed to suggest, was not just the ability to be free to sit or wander. To be semi-wild was to have a regimented schedule similar to a workday for human wage laborers. The joke, of course, lay in the content of that schedule: sleeping, sitting, wandering. Yet it nevertheless emphasized the ways in which orangutans were disciplined by the human time constraints of wage labor. The wage they earned was simply their keep. To stay alive, they had to cari makan. If we follow Ricky's joke about orangutans having a job at Lundu Wildlife Center, their compensation was food. They literally had to cari makan.

Food for orangutans under these conditions was as much a manifestation of industrialization as was Ricky's punchline that orangutans have jobs. Orangutans in the wild eat an assortment of over two hundred plants (Galdikas 1988). Primatologists estimate that 40 percent of their energy is expended through securing food (Rodman and Cant 1984). In times of scarcity, which come in regular seasonal intervals, they are known to survive on tree bark (Campbell-Smith et al. 2011).

Orangutans at Sarawak's wildlife centers are given food support twice a day, regardless of whether or not they were able to forage in the forest. At the time Ricky and I spoke, Lundu Wildlife Center held seven adult and juvenile orangutans in six iron-barred cells within the orangutan house at night and four exterior exhibit spaces during the day. Feeding entailed workers and helpers leaving food out in the exterior enclosures as a way to entice the orangutans to leave their night cages. Once they were out in their exhibit spaces, their cells were cleaned by cage workers and helpers—commercial volunteers who paid thousands of dollars to perform manual labor (Parreñas 2012). For the juvenile orangutans undergoing jungle skills training, a worker would take them out to the forest by either guiding them to walk, piggybacking them, or holding them at the hip. Once in the forest, those young orangutans had a chance to explore, play, and wander through the canopy until summoned back to eat and return to their cages in time for a lunch break. After lunch, they were either taken back to the forest or—if the workers had to carry out a labor-intensive job, which often was the case—the orangutans would stay in their cages until the next day. By 4:00 p.m., all the orangutans would be brought back in, their second meal waiting for them, sometimes along with a parcel made by a volunteer. The contents of the banana leaf-wrapped parcel were often fruits or old plastic water bottles with sand in them. This was supposed to mentally stimulate them and fight the boredom of captivity. This schedule remained consistent throughout seasonal changes, but it broke down when orangutans got free, which they sometimes did, and also when the younger orangutans were brought out to the forest for a week of jungle skills training.

At the other site in Sarawak, the Batu Wildlife Center, orangutans are also fed twice a day on public platforms while being photographed by hundreds of tourists. (The workers' job there is to mediate the risky space between tourists eager to get close and freely roaming orangutans who are so acclimated to humans that they are not afraid of them. Some orangutans are notorious for biting people, and such bites require serious medical attention.) Even though the orangutans hypothetically have the ability to forage, and they especially do during the wet season, there is not enough wild food to sustain their population. Hence, they are given food sources throughout the year.

Like industrial workers, semi-wild orangutans at Lundu Wildlife Center are compelled to subsist on white bread. Even during the fruiting season, an ordinary meal fed to an orangutan would sometimes include two slices of white bread along with fresh seasonal produce like bananas, guavas, sugarcane, or papaya. All of these fruits are regional cultivars—some introduced through Spanish and Portuguese trade from across the Pacific (guava, papaya) and others domesticated in Southeast Asia (sugarcane, bananas) (Kennedy 2008; Keegan and Carlson 2008; VanBuren et al. 2015). In American zoos, an orangutan diet consists primarily of animal feed like Purina Monkey Chow or dry or canned ZuPreem Primate Diet, which include soy, corn, wheat, and animal fat (Barbiers 1985). Semi-wild orangutans' food is neither wild nor is it completely produced through agricultural industry like it is for zoo captives. Similar to their status as "semi-wild," the food they earn is semi-cultivated.

Orangutan mothers in the wild introduce their infants to a wide-ranging palate that includes hundreds of species found in the forests in which orangutans dwell. Thus, it may come as little surprise when semi-wild orangutans forage for foods beyond the cultivars of their assigned diet. One memorable instance was when center staff members Layang, Apai Len ("Apai" is a respectful appellation of "grandfather"), and their co-workers were training young orangutans in the forest. Ching the orangutan must have seen them, because on their way back to the ranger's station where they were spending a week for the juvenile orangutans' overnight jungle skills sessions, they saw Ching inside the hut. Layang stormed toward the hut, and then Ching took off, leaving behind a trail of strewn about and crunched up instant noodles, cookies, crackers, rice, and instant coffee. This plunged the work crew into a crisis: there was no way they could remain there for the week without food.

Layang, the fastest and most able-bodied of the group, offered to trek down to the main camp in order to replenish their supplies. Before he left, he suggested to the remaining party that they boil a pot of water and throw the hot water at Ching if necessary. The party included two elderly workers, Apai Len and Apai Julai, and their young and college-educated boss, the junior officer Cindy, who was usually kept busy in the park headquarters with administrative tasks. It was Cindy's first overnight jungle skills session. While waiting, she passed the time by slipping food through the floorboards to the four-year-old orangutan, Gas. At one point, Cindy noticed that Gas's eyes seemed to look different. She stopped. Suddenly, much larger hands than Gas's pushed through the planks: Ching broke through the floorboards.

Apai Len greeted Ching by dousing her with the pot of boiled water. But by then, the water had cooled to room temperature. Everybody ran in fear of Ching's response: not only Cindy and Apai Len and Apai Julai but also the younger orangutans. Once Cindy opened the door to get out, the juvenile orangutan Lisbet jumped on Cindy and bit her on the hand, impairing her writing and typing for the ensuing months. The chase stopped when Apai Julai struck Ching with a floorboard.

The point of this strange story for my purposes here is that orangutans are not just driven by a mechanical hunger. Semi-wild orangutans cannot simply be disciplined with food as a wage and as encouragement for their workplace performance of wandering, sitting, and eating. Here, we see orangutans driven by something other than simply the satisfaction of an appetite.

I am reminded of Rosaldo's question to fellow anthropologists: Why "bother getting up in the morning and face nothing but routine drudgery?" (1986, 102). Humans are not the only ones subject to boredom. And Ching, eating the same raw sweet potatoes and a rotating set of four to seven other cultivars instead of the two hundred or more species her very recent ancestors ate, very likely tastes boredom with every bite of her semi-cultivated diet. Regardless of our inability to read her mind, we outsiders certainly can see that this eventful moment broke the monotony of workaday life at Lundu Wildlife Center.

Finding Food instead of Work

The distinction between being a wage laborer and an agrarian laborer is blurry for people like Apai Julai and Apai Len as well as the couple with whom I stayed, Kakak and Ren, who were respectively in their thirties and forties. This was not just because they were Iban, famed in anthropological literature as shifting cultivators (Freeman 1955; Kedit 1980; Ngidang 2003). All four refused to work for wages during the rice-harvesting season, which spans a couple of weeks of the year. Even though they lived in the outskirts of the city, they rented paddy fields to cultivate rice for their household, driving for about an hour away from the city. As newer generations engaged in urban employment sectors, a division of labor began to fall along age lines: parents and grandparents cultivated rice while children and grandchildren engaged in wage labor. Buying rice was not only expensive, it could never taste as good as the varieties personally selected and grown. Apai Julai's family preferred black rice, for instance, while Kakak and Ren's household preferred a mix of red and white.

Attesting to the social significance of growing one's own rice, Apai Len and I had a conversation in which he tried to place me in this rice-centered social world. I had answered in the affirmative when he asked if my family ate rice. He then asked if my parents grew their own rice. The idea was unfathomable, since at the time my parents lived in a notoriously postindustrial and economically depressed American city with a dry chaparral landscape. He then asked if my grandparents in the Philippines had grown their own rice. The region from which they came was long dominated by a planter class producing sugar for the world market by exploiting an oppressive debt-bondage system since the mid-nineteenth century (Kaur 2004; McCoy 1982). My grandparents in the Visayas region of the Philippines did not cultivate their own rice, which at that moment felt shameful. To labor for food, to cari makan, is to have proximity between the uncertainty of subsistence farming and the precarity of wage labor in a subcontractual system. Agrarian labor and wage labor did not have the same proximity for me as a Filipina from a bankrupt American city but whose Visayan mother came from a former landowning family as it did for Apai Len as an Iban Sarawakian who has to cari makan.

The work to be done at Lundu Wildlife Center was physically demanding. Apai Len's co-worker Apai Julai knew this well. He had been bitten by residents on three different occasions. On one occasion he was bitten by a macaque. On another, it was a gibbon. The most recent time was the orangutan Ching. This was in the aftermath of him having hit Ching with the floorboard plank. He described Ching's reasons as rooted in her feeling of *sakit hati*, a feeling of being so wronged that it hurts the guts. One day, months later, he was caught in the rain and waited it out on the viewers' platform overlooking the captive orangutans' enclosures. She approached him at the stairs, which was the only entrance and exit. He was trapped. She bit him, scraping her teeth against his forehead so that he had a bloody gash.

This compelled Cindy's fellow junior officer Lin to photograph Apai Julai's injury. Lin was afraid of Ching, who had a reputation for biting women, and thus Lin was against Ching being freely mobile within the confines of the park. Lin circumvented her immediate boss who supported Ching's physical independence and complained to the regional director, who then ordered the park to force Ching and her baby back to captivity.

Ricky, Layang, and their co-workers disagreed with the work order, since they strongly felt that the orangutans had to be free if this site were to work as an orangutan rehabilitation center. After Lin took photographs of Apai Julai's wound, she sent him home. A rumor then circulated that Apai Julai, instead of going home, went to his plot of land and worked on his pepper plants. Whether or not the rumor was true, Apai Julai's ability to perform agrarian labor then became proof that he was not truly injured: if he could perform agrarian labor, he could perform wage labor working with orangutans. Thus, the fear of either life-threatening or debilitating risk with having Ching unrestrained, fear that Lin and the regional director shared, had insufficient cause. Apai Julai's ability to work on his field of pepper plants meant that Ching's return to captivity ultimately lacked justification.

To work was to cari makan, to find food, to earn one's keep, and to simply subsist. Subsistence seemed to only be feasible if mediated through the regimented hours of the workday, in the forced sociality of a workplace, and in the constrained space of the private-public partnership orchestrated by the officers of the Forestry Corporation. This was the case for both workers like Apai Julai and for displaced orangutans like Ching.

Substituting Subcontractual Labor

Looking back at Bidai's situation as a new worker inducted into a twenty-four-hour workplace, we see that Bidai had to cut or constrain social ties. However, in the current era, it seems that human social relations are not forsaken but are rather open to exploitation. This is different from Barbara Harrisson and her decision to take on responsibilities at the Sarawak Museum as the wife of the museum's curator, including raising displaced orangutans sent to the museum. It also differs from the bonds of kinship, labor, and obligation that Sylvia Yanagisako (2002) illustrates in her study of family businesses in Italy. What I suggest is that for wage laborers struggling to get food on the table, the obligation is even more pressing. In this respect, these relations are more conceptually aligned with the family relations of Cambodian Americans described by Aihwa Ong (2003) who see their family members as free sources of labor to exploit.

The wildlife center is not a family business but is run by a corporation. Most of its staff members are subcontracted workers. Following Anna Tsing's (2009) argument that contemporary capitalism and its supply chains exploit and mobilize gender, race, class, and culture—which are often seen as superfluous or irrelevant to capitalism—I see that the work of caring for wildlife and running the center in this moment of cheap subcontracted labor also exploits and mobilizes family ties, in particular gendered obligations women and girls may feel toward the men in their lives.

For example, Kakak worked as the orangutan nanny. Before that, she was an unofficial substitute for her husband, Ren, an animal keeper. After the Forestry Corporation took over park and biodiversity conservation responsibilities from the Department of Forestry, he no longer had access to overtime pay, which was important for their subsistence. So he and Kakak had to become flexible when it came to getting food on the table for their family. On days he was able to get a better-paying construction job, Kakak substituted for her husband at the Lundu Wildlife Center. At first, she hesitated. She was scared of the job. She was scared of facing the crocodiles, to which she'd have to toss dead chickens across a wire fence. She was scared of facing the binturongs, black with yellow eyes, coarse fur, and a laugh that sounded to her like it belonged to something evil. But her husband said she had to do it because they needed the money. So she did.

Like Kakak and Ren, Adik and her father lived in the same longhouse as Apai Julai. Adik was perhaps ten or eleven years old. Similar to Kakak stepping in for her husband, Adik stepped in for her father when he was too drunk to show up for work. Her father worked as the overnight security guard who watched the gate, unlocking and opening it for those who had permission to enter the park, which was limited to workers and volunteers living in staff quarters. When she once opened the gate for Wilson, a Bidayuh staff member, and myself, Wilson asked her with suspicion where her father was. She answered with a scowl that he was bathing. Her father was eventually fired. I am not sure how she and her father put food on their table henceforth.

Many months later, when I moved into the longhouse, I gifted Adik pencils. She later shared cacao fruit with her peers and me. The price of cacao was so low, everyone with cacao trees at the longhouse let them go feral—or semi-wild. I never had cacao like that before, and she taught me how to eat it. Such a small gesture hints that not all cultivars are to be treated as commodities, even when the ability to "find food," or cari makan, is insecure.

Years after Kakak's initial lessons in Lundu Wildlife Center's workplace demands, Kakak could joke about it. Over drinks with her in-laws, my friends from Kuching, and me, she laughed boisterously as she imitated binturong sounds in an attempt to convey why they were so scary. I shared her laughter, getting the tragicomic joke that even the most frightening of sensations was worth overcoming for a paycheck.

I hope years from now Adik can laugh about the time she worked as a security guard on behalf of her father. Adik in particular shows that subsistence is more than "trying to feed their families and keep their children alive" (Nash 1994). It is also about finding sustenance through laughter, sociality, and the fruits of feral and semi-wild conditions.

Conclusions

Through their jokes, Ricky and the datuk find humor in the serious business of orangutan rehabilitation. Orangutan rehabilitation is not just about responding to the threat of species extinction. It is also about semi-cultivated, semi-regulated, semi-feral, and semi-wild possibilities.

When we look at the experience of the orangutans on site, we see the uniquely ambiguous space offered by the wildlife center. It is not wild where the orangutans are free to roam and free to starve. Nor is it a zoo where the orangutans live off agricultural industrial products in a state of perpetual forced dependency, where their lives are only possible because of human intervention. It is a space of human-animal relations where each side of the interface has to experience vulnerability with the other, as Apai Julai's experience of three different bites exemplifies.

When we look at the experience of the people on site, we see how the work is clearly demanding, and yet what constitutes a worker is not clearly delineated in respect to time or job title. We see how, in the 1960s, workers were expected to work long hours and give their personal interests, leisure, and chosen social relations lower priority. In the 2000s, we see that the question of who counts as a worker in a subcontracted system gets confounded when we see female kin relations securing their male relatives' wages by stepping in as their substitutes.

The experiences for orangutans on one hand and people on the other are not just parallel stories of how to make a living in a space of loss, displacement, and massive transformation. They are bound to one another. We see this in the regimented hours of the day, well into the night, with the demand that everyone and every body live by the clock. We sense this in the push they each feel to creatively find sustenance, whether that is raiding a ranger station or resorting to substituting for one's father when he is too drunk to work. Yet life here is not just about bare survival. This is not a world of ZuPreem Primate Diet or Purina Monkey Chow. What we see instead is that working for money and sustenance is one action among many. These different actions include using the state agency's motorboat to break from the isolation of the workplace and attend a party in the closest village, two hours away; biting the human hand that feeds you; and temporarily refusing waged work to harvest rice. These actions all have uncertainty tied to them, whether that is in the risk of drowning, the unknown consequences of an interspecies workplace, or what to do after one's single father is laid off. It's the very uncertainty that undermines a mechanical idea of nature working. Work is not a given. It is not something merely in the environment that is ripe for extraction and exploitation. Indeed, the uncertainty of what work is worthy of one's effort for gaining compensation sustains the ongoing joke of what work is worth a given wage.

Note

1. Cari makan is pronounced as "chari makan."
2. Thanks to Patrick Daly for identifying Barbara Harrisson's handwriting.

CHAPTER FIVE

The Heat of Work

Dissipation, Solidarity, and Kidney Disease in Nicaragua

ALEX NADING

In both urban and rural workplaces, exposure to heat shapes the experiences of people living with chronic kidney disease (CKD). CKD is perhaps most associated with diabetes. In diabetics, heat exposure exacerbates CKD's effects, speeding dehydration and straining the kidneys' already diminished capacity to filter and excrete metabolic waste. But diabetics are not the only ones affected by CKD. Around 2000, Nicaragua was among the first places in the world where a new form of CKD, known as "chronic kidney disease of nontraditional causes" (CKDnt), was detected. CKDnt is not related to diabetes. It occurs in otherwise healthy people, mostly men. In Nicaragua, the majority of these men work on sugarcane plantations. Whereas heat exacerbates CKD's effects in diabetics, evidence suggests that heat may actually help cause CKDnt in sugarcane workers (Roncal-Jimenez et al. 2016).

CKD is a condition in which the vulnerability of the body seems conjoined with that of the biosphere. It is a condition in which the effects of overcrowding, over-farming, and overexertion reverberate and resonate between persons and working environments. CKD is, among other things, a thermodynamic condition: a dysbiosis between a system and its surrounding context. It is a disease in which working bodies can no longer adapt to atmospheres, yet it is also one in which atmospheres seem oversaturated with working bodies. CKD, in both its "traditional" and "nontraditional" forms, is a disease of exhaustion. It is thus an apt place to begin charting the limits of work on a warming planet.

Heat is rarely what work aims to produce. Rather, it is a material signifier of work. Whether in the kitchen where food is prepared or in the field where food is harvested, heat, measured in temperature and sensed by bodies, makes workers, managers, engineers, and occupational health specialists aware of work's limits, its wastes, and its excesses. These manifest themselves in sweaty brows, burned skins, desiccated soils, and delirious minds. As the stories of Nicaraguans living with CKD illustrate, however, heat is not just a byproduct of working life. Power, motion, and resistance—seemingly the categorical opposites of waste and excess—all also manifest themselves in heat. We measure energy by the calorie, just as we metaphorize the strength of social movements (including labor movements) with terms like "boiling," "burning," and "seething." Measurement and feeling combust, politically and physically, when the heat of work becomes unbearable. Laborers often make their mark in the world through intentional acts of burning factories or sabotaging machinery and causing it to overheat. In this way, heat is less the objectification of the labor relations that connect humans to one another (and to nature) than a generative context for those relations. Heat allows workers to think relationally, both in work and about work, before those relations are reified or encapsulated in a commodity or object.

Below I compare two stories of how heat and work intersect in the lives of Nicaraguans. One is the case of a diabetic woman with CKD in Ciudad Sandino, a crowded, economically marginal area of peri-urban Managua. The other is the case of sugarcane cutters with CKDnt in western Nicaragua's sugarcane corridor, the country's most intensive agricultural area and one of its most lucrative. Taken together, these cases show how the heat of work catalyzes not just suffering but also solidarity, in the form of kinship, wage labor, and transnational labor politics. As the stories in this chapter show, heat can stress and even destroy class, political, and kinship bonds, but heat can also catalyze fleeting forms of togetherness. In an era of catastrophic environmental and economic upheaval, the basis of solidarity may lie less in material groundings than in thermodynamic interactions.

Case 1: Summer in the City

"*Que calor.*"

"This heat," doña Julia said, sighing, as she rocked back and forth in her chair, not looking up from the half-sorted pile of black beans at her kitchen table. Beads of sweat glistened from her brow to her freckled forearms to her swollen ankles.

"Que calor" was often her way of greeting me when I entered her small house, located on one of the main streets in Ciudad Sandino. Tucked between an evangelical Christian church and a hardware store, doña Julia's house seemed, even in this persistently scorching city, especially hot. Like most every other house in Ciudad Sandino, doña Julia's sat on a ten-by-thirty-meter lot, fronted by a concrete wall topped with wrought-iron bars. From its front porch to its rear latrine, the house was capped by a sheet metal roof. The only break in the cover was a two-square-meter opening that permitted the upper branches of an emaciated mango tree to escape into the sunlight. When it rained, the small patch of dirt where the tree trunk met the concrete floor would be fitted with a plastic bucket to collect the water that cascaded from above.

But it hadn't been raining much. We were in the midst of what Nicaraguans call the Veranillo de San Juan. Falling around the time of the June solstice, after the first weeks of the rainy season have alleviated the stifling hot and dry of summer (*verano*) (roughly February through April), the Veranillo de San Juan, as its name indicates, is a time when summer makes an encore appearance, undoing any cooling effects of those initial showers.

This relapse felt cruel, violent even, especially for doña Julia. Like many fifty-something women in this part of Nicaragua, she suffered from *azúcar* (sugar). She was a type 2 diabetic, and one of thousands of Nicaraguan diabetics whose disease had led to CKD. She was able to manage and monitor her condition better than many others, thanks to a rich network of decently employed and caring kin. Still, doña Julia was prone to exhaustion. She only occasionally ventured out. One of her several granddaughters or nieces was nearly always with her, helping attend customers at the small *pulpería* in the front of the house where she sold ice, rice, beans, and an array of cheap, sugary snacks, including packaged candies and her specialty, *chocobananos*, frozen bananas rolled in confectionary chocolate. Doña Julia's daily life, like those of the plantation workers I discuss below, was beset by an interplay of sugar and heat. In its refined form, azúcar was her stock in trade. In the form of blood glucose, azúcar was her nemesis. But to gain both its maximum market value and its maximum potency as a diabetic danger, azúcar had to be compounded with heat.

It was perhaps because she ventured out so infrequently that doña Julia had insisted that her husband, don Luis, enclose the entire house—save the little mango tree—in sheet metal. Most other houses in urban Nicaragua include an open-air patio where trees and other plants break the monotony of gray concrete or volcanic black dirt. Some of these plants, including tomatoes, mint, and spinach, have well-known medicinal qualities (Shillington 2013). Indeed, several are cultivated by Nicaraguans to combat the symptoms of diabetes, including fatigue. Don Luis's enclosure of the house frequently made its indoor temperature feel higher than the temperature outdoors, despite the shade. But doña Julia insisted that she would have it no other way.

"¡*Cierra la puerta!* [Close the door!] There are *pandilleros* [youth gang members] out there," she would scold her grandchildren and me from the kitchen table, almost as soon as she heard the creak of the gate marking someone's entrance. "They will rob you and cut you! Be careful!"

Doña Julia's warnings about pandilleros came in hard staccato notes, a gravelly pitched voice slicing through the din of passing buses and noisy church organs. When she was not picking through beans or preparing chocobananos, she could be found with her plastic chair positioned in the doorway between the kitchen and the small sitting room, where the television, tuned to the news, broadcasted the latest stories of street crime.

In doña Julia's house, the force of heat was compounded by climate patterns, hardware, mass media, and metabolic disease. Heat trapped doña Julia inside, and in turn, the construction of her house trapped heat with her, waves of sunlight caught in a convection current with the embers of the wood fire and the output from the freezers and fridges that contained ice, Coke, and chocobananos ready for sale. Doña Julia's warnings about keeping the door shut came as alarm blasts. Her equally frequent reminders about the heat, however, came in resigned, almost plaintive moans.

Que calor.

Case 2: A Mystery Epidemic in the Sugarcane Zone

Jorge and Ulises Pacheco are brothers and former sugarcane cutters. Both of them live in the town of Chichigalpa, in northwest Nicaragua, and both have been diagnosed with CKDnt. CKDnt was virtually unknown to medical science until the early 2000s, when thousands of Central American sugar plantation workers, most of them men in their thirties and forties, and none of them diabetics, began dying. The scale of the epidemic in Nicaragua is staggering. Between 2005 and 2009, the number of deaths from kidney disease among men in the country rose by some 41 percent, mostly among cane cutters.

Jorge and Ulises, like many other sugarcane cutters in Chichigalpa, learned about their condition at the start of the cutting season. At the beginning of the harvest each year, cane cutters seeking work on plantations owned or operated by the local mill, known as the Ingenio San Antonio, had to report to the company's private clinic, where they would undergo blood and urine tests designed to determine their fitness for work. During the late 1990s and early 2000s, Jorge, Ulises, and hundreds of other would-be workers, some with more than twenty years of experience cutting cane, were dismissed from the rolls. Tests on these men's blood serum revealed unusually high concentrations of creatinine. Creatinine is a waste protein secreted during muscle metabolism and, under normal circumstances, filtered out of the body by the kidney. Elevated creatinine levels are a biomarker for compromised kidney function. Over a course of sometimes months, sometimes years, these men's kidneys will stop working. Dialysis can stave off death for a while, but CKDnt is in nearly all cases a terminal diagnosis.

When the Ingenio San Antonio fired Jorge, Ulises, and other cane cutters for having elevated creatinine levels, it shrewdly indemnified itself against any legal obligation to provide them with medical care. Had Jorge, Ulises, or their comrades been injured on the job, they would have been entitled to disability insurance benefits through the national social security scheme, but their kidney failure, according to the Ingenio, was not a problem of labor conditions but of laboring discipline. Even though the dismissed workers were not diabetics, authorities insisted that some combination of lifestyle and genetic factors must be to blame. Perhaps they drank too much alcohol after work or consumed too many sugary drinks to beat the heat while on the job. Perhaps they took too many nonsteroidal anti-inflammatory drugs (NSAIDs) to dull the pain of overexertion (high doses of NSAIDs like ibuprofen can compromise kidney function).

Jorge, Ulises, and other workers were not convinced. When they contemplated the mysterious new syndrome, they wondered about the long-term effects of chemical pesticides, about the dust, and about the difficulty of eating healthy meals amid the daily demands of cane cutting. They also wondered about the heat.

Work in the cane has always been deadly, exploitative, and hot (Mintz 1960). Indeed, a blast of heat marks the start of the cutting season, as grasses are burned away to expose the sugary stalks hidden underneath. No one I know in Nicaragua envies the job of cutting the massive, sharp stalks in triple-digit heat, day after day, but in northwest Nicaragua, there are few jobs as reliable. Occupational health experts compare harvesting sugarcane to running a half marathon in hundred-degree weather, going home and going to sleep, and doing the same thing again — six days in a row. The toll, in terms of body temperature and dehydration, is massive, and the kidneys bear the brunt.

Since their firing, Jorge, Ulises, and others have begun to contest their dismissal by the Ingenio San Antonio. In coordination with Nicaraguan and international labor and environmental activists, they have mobilized to turn CKDnt into a basis for addressing the inequities of global sugarcane production. Amid the epidemic, heat has thus catalyzed a new form of solidarity. It has invited activists, scientists, and others into the "troubled ecology" of Nicaragua's sugar corridor (Besky and Blanchette, this volume).

Heatscapes and Workscapes

Heat has long been at the heart of economic and social theories about what it means to work and, just as importantly, what it means for persons or objects or animals to resist or refuse work. During the early days of industrial capitalism, the capture of heat in steam engines helped push the limits of mechanical production, even as it reaffirmed those of biological reproduction. Machines did not save labor so much as they redirected energy. As Anson Rabinbach (1992) notes, the very concept of "fatigue" as a dissipation of labor-power—whether of human bodies or of machines—emerged in the nineteenth century through the sciences of economics, social medicine, and mechanics. Early industrial machinery released tremendous, unprecedented amounts of heat not only into turbines or pistons but also onto the surrounding factory floor, into the atmosphere, and into the bodies of laborers. Exhaustion was an economic, environmental, and social problem. For the labor movements that sprang up around this time, what a working body could not take any longer was not just the humiliation but the heat. Refusal and reduction of work (see Dave, this volume) are—and remain—as much about confronting physical conditions as class positions. If one takes the refusal to work as an organizing principle for politics, for example, one rejects the notion that one's value as a person or political actor stems from one's ability to "take the heat" of the factory, or the office, or the hearth (see Weeks 2011).

Heat thus shapes what Thomas G. Andrews (2008) calls the "workscapes" of capitalist life. In her house, doña Julia created a sugary, superheated workscape in which the motors of her refrigerators, the convection of the roof, and the wood on the fire all mediated the metabolism of sugar in overlapping ways. Sugar and heat were the twin engines of the family business and the family matriarch. It is not too difficult to build an analogy between those overheated factories, staffed by fatigued bodies, and the overheated household.

The sugar plantation is also such a workscape. In an account of life in and around the Ingenio San Antonio—the very sugarcane mill that fired those two brothers, Jorge and Ulises—historian Jeffery L. Gould details how, back in the 1920s, workers and their neighbors had a singular explanation of the origins of the plantation's power. According to a myth that circulated among the workers, "the company had signed a pact with the devil in order to further accumulate wealth. . . . The devil pact specifically allowed the company to convert dead laborers and their families into cattle" (Gould 1990, 29). The company would then sell these cattle to a hacienda. "The death of [a laborer] did not mean the end of his service to the company," Gould writes. "Rather . . . the worker continued to produce wealth for the company . . . either as oxen or as food for the work force" (30).

Ruthless efficiency, indeed.

The analytical and political challenge in both the urban house and on the plantation is to tease apart those layers and vectors of heat: to make sense of overlapping currents and waves. Consider the linkage between the manner in which sugarcane workers (and oxen) produce wealth and the manner in which foods like chocobananos and meat produce energy. That linkage can best be understood as thermo-economic. Economically, heat mediates between composition and decomposition, production and decay, social solidarity and social division. In anthropology, this mediating role once made heat into a key theoretical anchor. Consider, for example, the energetic ecological theories of early environmental anthropologists, who tended to measure the flows of caloric (i.e., heat) energy through populations. Roy A. Rappaport (1968) saw the consumption of mass quantities of pork among the Maring of highland Papua New Guinea as both an infusion of calories and as a catalyst for ritual solidarity among humans, animals, and ancestors. Similarly, Marvin Harris (1966) saw the energetic power of uneaten cattle in India as central to their sacred status.

Those Nicaraguan peasants in the 1920s probably weren't thinking about the relative caloric value of cattle and sugar, and they certainly couldn't have envisioned the present global crisis, but the entanglement of heat and economy was central to the formation of industrial capitalism and wage labor. The power of capital to turn the cold bodies of the dead into warm-blooded cattle must have struck fear into the hearts of Nicaraguan peasants, in part, because it indicated not only a vital capacity to extend labor-power into the afterlife but also a capacity to make life—seen as physiological heat—coincide with work.

Karl Marx famously called factory machinery "dead labour, that, vampire-like, only lives by sucking living labour, and lives the more, the more labour it sucks" (1976, 342; see also Hetherington, this volume). Capital, he realized, had to brutally convert the vital heat of the warm body into the physical heat of the machine in order to extract surplus value. If you're the manager of a factory or a plantation, you end up being very concerned with the sleep habits, sexual appetites, and, of course, food intake of your workers, as indeed Jorge and Ulises's bosses were (see Besky, this volume). In the colonial plantation economy of Latin America and the Caribbean, sugar—what Sidney W. Mintz (1985) called a "proletarian hunger-killer"—became a mechanism for doing this. Sugar carried cheap calories—the kind of calories found in doña Julia's chocobananos—but it also carried a latent contradiction. The acceleration of capitalist accumulation (and sugar consumption) over the decades of the nineteenth and the twentieth centuries fueled the slow-moving metabolic epidemics of diabetes and obesity. Of course, if you're running a small home business and trying to keep a household full of children and grandchildren fed, clothed, and safe from pandilleros, you have similar concerns.

In Nicaraguan workscapes, heat is as close to a total social fact as anything else. It courses through climatic, economic, architectural, biological, and labor processes. Women like doña Julia have fine-tuned methods of differentiating the hundred-degree heat of the Veranillo de San Juan from the ninety-three-degree heat of a "winter" day in November. Heat shapes Nicaraguan workscapes not only in a phenomenological sense or in a homeostatic sense (one could certainly follow Rappaport and Harris into these workscapes and make a functionalist argument about energetic imbalance) but also in a more dynamic, entangled sense. The heat in Nicaraguan workscapes is not just matter giving rise to symbolic orders; it is a material force and a semiotic force all at once. Jorge and Ulises's work in the cane fields, before they were fired, channeled heat from sun and stalks into bodies and motors. Human cane cutters are thus not the only ones in the sugar corridor who are working. They are surrounded by rows of energy-packed sugarcane: caloric stores that might be burned today as food by human bodies and tomorrow as ethanol by internal combustion engines. The pesticides they apply to the cane and the microbes in the soil are all also working: they harness and transfer energy by photosynthesizing, metabolizing, and intoxicating (see Kim, this volume). When sugarcane workers like Jorge and Ulises began to mobilize to address CKDnt, as I show below, it was the immanence of heat—seemingly everywhere yet nowhere in particular—that invited scientists, activists, and others into this workscape.

Likewise, when doña Julia uttered her exhausted "que calor," she was not so much pointing to an invading, inhuman outside force as she was inviting her children, me, and anyone else who would listen into a workscape. Doña Julia's work as a cook and as a head of household channeled heat through domestic and somatospheric space. Her refrigerators, her wood stove, the sunlight converted by metal sheets into convection waves, and the glow of the television news were both what sustained her and what exhausted her. The words "que calor," "this heat," were an attempt to provincialize and domesticate a thermodynamic problem. They were an attempt to tend to the affective gap between measuring and feeling.

This view of heat, as an immanent presence in work, rather than the input or output of work, is quite at odds with the view that predominates in economics, where heat is often seen as an externality: as the "free energy" that comes from the sun or the earth or as the waste produced by inefficient machinery. Medical anthropologists have shown how this economics of externality leads to uneven forms of suffering. The laborers who tend to be exposed to extreme heat are often already subject to racialized or gendered discrimination. Latino farmworkers in the United States, for example, are not exposing themselves to extreme temperatures out of ignorance or poor choices but out of economic necessity. If those workers claim (falsely) that they have a congenital tolerance for heat—that as African Americans, Latinos, or as Indigenous people, they have an inborn ability to spend longer hours in higher temperatures—this must be in part because they need to make such a claim in order to bridge that gap between measuring and feeling. Such claims—as affective as they are objective—make them both physically available to do strenuous work and psychologically prepared to endure the physical hardships that such work inevitably imposes (Horton 2016; Holmes 2013; Farmer 1999). Based as it is on false assumptions about the body's capacity, the work ethic of the farm laborer becomes a form of structural violence. Suffering—and the human response to it—is never a matter simply of metrics. Neither is heat.

Heat is not simply "out there" to be measured or experienced; it does not cause good or bad health, nor is its presence or absence merely an effect of good or bad social relations. Heat as measured temperature and heat as embodied experience are not two sides of a coin; rather, the measurement and experience of heat are always entangled (Barad 2007). Heat does not so much *cause* work to begin or end as it *catalyzes* working relationships. Heat makes work into something that people can simultaneously evaluate, endure, and resist.

Energizing and Exhausting

When I first met him in 2015, Jorge recalled his initial theory about what caused the rash of CKDnt deaths in the sugarcane zone. For decades, the little town of Chichigalpa was surrounded by a diverse agricultural landscape. Jorge explained that when he was a child, the land in this area would have sat under sugar, but cotton was the dominant commodity crop. In the villages that dotted Nicaragua's northwest coast, farmers in his parents' generation held on to small subsistence plots where they could cultivate vegetables, fruits, and a few cattle (Gould 1990). It was not until the 1980s, he said, that sugar really began to dominate the landscape. In part, this transition was due to the repeated failure of the cotton harvest, resulting from the aggressive use of agrochemicals, promoted by Nicaragua's government in the 1960s and 1970s as a development strategy, as well as a glut in the global cotton market (Murray 1994). Sugar was resilient, and the corporation that owned the Ingenio San Antonio, the Pellas Group, was the most powerful in Nicaragua. Its signature product, Flor de Caña rum, bears a picture of the Ingenio on its label, with the massive volcano San Cristóbal—its own kind of heat source—looming in the background.

Over the course of the 1980s and 1990s, even as Nicaragua passed through a decade of revolutionary and counterrevolutionary war and reconstruction, the Pellas Group's sugar concern, Nicaragua Sugar Estates Limited (NSEL), either bought or leased more land for sugarcane production. Despite the political turmoil, business was good. Small farmers in the villages began feeling pressure to capitalize on the sugar expansion, and as their houses became increasingly enveloped in cane, they occasionally found their cattle dead from pesticide poisoning.

What Jorge was describing was the growth of a monoculture (see Besky, this volume). By the end of the 1990s, sugar's dominance over the landscape was more or less complete. New pesticides had taken over for the old ones, and as the cane crept closer to the villages, so did the chemicals. Workers and their Nicaraguan and international supporters all agreed that the growth of this monoculture was taking a toll. Well before the Ingenio San Antonio began dismissing workers with compromised kidney function, residents had been communicating their concerns about the loss of crop diversity and the drying of wells—as well as the loss of cattle due to pesticide poisoning—to American and European development workers, labor activists, and humanitarian groups, as well as to relatives and friends in León and Managua. In Jorge's account, concerns about land tenure loss and soil and water contamination preceded concerns about the epidemic of deteriorating kidney function.

But it was the Ingenio San Antonio's brazen dismissal of workers with CKDnt that turned this concern into activism. By 2005, the epidemic was well undeway, killing hundreds of men each year. Fear was mounting that if the workers did nothing, the problem would get worse. The rum business was steadily growing, but NSEL had larger ambitions.

In 2006, amid a worldwide biofuel boom, the World Bank's International Finance Corporation (IFC) was preparing a $55 million loan for NSEL. The loan would enable the company to expand its plantations and construct an ethanol plant near Chichigalpa. Alerted to an obscure rule in the IFC loan procedure, Jorge, Ulises, and other workers who had been fired after showing signs of renal failure began working with students, lawyers, and activists from the United States to address the problem. In March 2008, before the loan was approved, the workers joined with the Washington, DC–based Center for International Environmental Law to file a grievance with the IFC's Compliance Advisor Ombudsman (CAO).

The workers and their lawyers amassed evidence indicating that the epidemic of kidney failure, the deaths of cattle, the poisoning of water, and a rash of respiratory epidemics were all correlated with NSEL's expansion of sugar plantations. In 2005, a study carried out by the Nicaraguan nongovernmental organization Profesionales para la Auditoria Social y Empresarial and the International Labor Rights Fund surveyed some 650 workers in the area. Eighty-three percent of the workers who participated in the study were between sixteen and twenty-eight years of age. As the study's authors noted, "The short working life of many of these . . . workers is related to the dehydration, malnutrition, and exposition — for long periods of time — to ultraviolet rays and agrochemical products. . . . These factors cause many chronic and serious illnesses. . . . These illnesses include skin cancer, kidney diseases, and sterility, among others" (PASE/ILRF 2005, 32).

To be clear, the CAO grievance listed a wide range of problems, from soil erosion to cattle die-offs to a rise in respiratory illness. But it was the story of the mysterious kidney disease and the summary dismissals that seemed to sway the World Bank's attention. In 2008, thanks to the grievance process, NSEL began negotiating a settlement with a workers' group called the Chichigalpa Association for Life (ASOCHIVIDA). The settlement included the appointment of a US-based research team to study what became known as CKDnt (Brooks and

McClean 2012). Amid the heat of the sugarcane zone, a new kind of solidarity was being formed.

Working with (and against) Heat

On the outskirts of Managua, a little under one hundred kilometers south of the sugarcane zone, doña Julia's experience of CKD, heat, and work was far less dramatic. Though she was once a militant in the Sandinista movement that overthrew Nicaragua's Anastasio Somoza dictatorship in the 1970s, by the time I met her in 2006, she was no activist. Like many chronically ill people in urban Nicaragua, she depended for her health (and the health of her kidneys) on the capacity and will of her younger kin to work. For many young men in Nicaragua, such work is to be found outside the country, in the booming construction industries of Costa Rica and Panama, Nicaragua's wealthier neighbors. While Nicaragua is an outlier among Central American countries in that it sends relatively few people north to the United States, its economy remains highly remittance dependent.

The economy of care for the failing kidneys of women like doña Julia, then, depends in no small measure on the continued availability of construction and other manual labor in the growing cities and tourist destinations that lie a few hot hours away by bus. The latent expectation in this labor market—as in the US agricultural market—is that some bodies are more tolerant of heat than others (Holmes 2013; Horton 2016). Nicaraguans are reputed in the more affluent parts of Costa Rica and Panama to possess a dogged willingness to work cheap and without complaint in difficult conditions. Beliefs about the resilience of certain working bodies underwrites the physical production of heat under capitalism.

In the workscapes of migrant labor, heat's contradictory guise as both a positive index of work-in-action and a dangerous index of work-in-excess becomes evident. Recent news from the world of heavy industry and construction indicates that the heat that has long driven the global economy might be its undoing. Consultants at Verisk Maplecroft (2015) have warned that rising temperatures could cut productivity in Southeast Asia by as much as 25 percent over the next thirty years. States, insurance companies, and labor unions from California to Texas are rewriting occupational health rules to deal with the threat of heat stress (Horton 2016).

On their return from temporary work trips abroad, doña Julia's children could rely on the warmth of home. The condensed workscape of the pulpería was indistinguishable from that of the household. Tortillas and beans would be on the hearth, and laundry could be washed and dried in the sun. Like many home cooks, doña Julia relied partly on wood fuel. Though gas cookers were cheap and readily available, wood was the only way to give certain homemade foods (*comidas caseras*) their proper taste. Smoke and fumes—the waste products resulting from converting wood into heat—conveyed flavor, but they also took their own toll on the household, much as the effluent from aerial fumigation and the burning of sugarcane fields at harvest time causes chronic respiratory distress up north in Chichigalpa. For doña Julia's family, heat was less an externality than an immanent presence in life and work, both exacerbating exhaustion and curing it.

Cutting Cane, Cutting Heat

When the US-based research team that was appointed after the settlement between ASOCHIVIDA and NSEL came to Chichigalpa, workers initially demanded that they test blood and water for the presence of pesticides. The researchers assented to this demand, but they also carried out more conventional epidemiological studies, trying to account for behavioral factors like diet and NSAID usage, and comparing kidney disease burdens in northwest Nicaragua to burdens elsewhere in the country (Ramirez-Rubio et al. 2013). Some investigations indicated that children in communities where CKDnt is prevalent might have been genetically predisposed to the disease (Laws 2015). Everything was on the table. Eventually, however, heat stress came to be the factor most strongly associated with CKDnt. Extreme heat exhaustion and heatstroke have always been known to compromise kidney function, but some investigators came to believe that a slower, less acute accumulation of heat stress was to blame for the epidemic in Nicaragua. As one group of authors punned, CKDnt in Central America had become "a hot issue" (Brooks, Ramirez-Rubio, and Amador 2012).

The heat was real and figurative. The support of the World Bank and of the Pellas Group for epidemiological research along with an agreement to provide medical care, restore insurance benefits, and supply bereavement benefits to workers and their families who had been dismissed by the Ingenio San Antonio were designed not only to find answers but also to heal the rift between industry and workers. To this end, the agreement has only been marginally successful.

Jorge Pacheco, one of the two brothers I introduced at the start of this chapter, is one of the leaders of ASOCHIVIDA, the organization that, through the CAO, negotiated the deal with NSEL. Just as the deal was being mediated, his brother, Ulises, formed a group that opposed allying with a plantation company. Ulises and many others rejected the premise that laborers could equitably negotiate with a corporation that had ignored laborers for more than a scorching century. This struggle between two dying brothers was over how to make justice out of heat.

Jorge and Ulises are no longer on speaking terms, but they are both working to mobilize national and international attention to CKDnt. Their movements have enlisted the help of food justice advocates, labor organizations, and global health actors. Transnational solidarity associations forged during the Nicaraguan Revolution (1979–1990) have been reconfigured as a younger generation of international activists has begun using the Internet and multimedia press releases to call attention to CKDnt. Some workers have formed splinter organizations and organized public protests and documentary films to link CKDnt to labor conditions. This approach has proved costly. After Nicaraguan police killed a CKDnt protester in 2014, many in Chichigalpa began to doubt the long-term efficacy of direct action.

Despite the turmoil, the activism that began in Nicaragua has turned CKDnt from a nearly unknown "mystery disease" into an issue that links global health, global climate change, and global labor struggles (R. Chatterjee 2016). It is possible that the kidneys of sugarcane cutters are responding to some kind of thermal tipping point. As a Colorado-based research team has suggested, conditions in the cane may have gotten just slightly worse—just hot enough—to spark a new pathology (Roncal-Jimenez et al. 2016). Temperatures in El Salvador, less than one hundred miles north of Chichigalpa, have risen by 0.5 degrees Celsius since 1980. The same studies indicate that work in the cane has also likely gotten more intense. Thanks to an increased demand for cheap sugar, hours are longer, and breaks for water, rest, and shade are less frequent. The heat that has always been there—as potential energy in sugar, as pounding sunlight, as surplus labor—is taking on a new embodied form.

These studies show how regular intervals of water, rest, and shade can stabilize creatinine levels and help maintain kidney function. At the same time, these studies are designed to probe the limits of working *with* heat. The only way to prove that work in the cane is dangerous to health is to run tests on people who work in the cane. Such studies aim to make plantation work more tolerable, not to do away with the plantation system. Across Central America, the prominence of sugarcane in the landscape—and heat in the workscape—is difficult to unthink (Andrews 2008).

The activism of Nicaraguan cane cutters has been largely responsible for bringing scientific attention to the sugarcane zone. Workers from Chichigalpa used international law, street protest, and alliances with epidemiologists and humanitarian groups to make their particular embodied experience of heat legible. But the most conclusive studies about heat and CKDnt have largely taken place outside Nicaragua. Inside Nicaragua, the government and the Pellas Group have successfully blocked nearly all attempts by researchers to perform longitudinal studies of heat in the cane fields. The state and the corporation remain able to externalize the problem.

Conclusions

The kidney has long been a site at which kin and social relations play out (Crowley-Matoka and Hamdy 2016). Brothers and sisters accompany one another over long periods of dialysis, and close kin can be among the most suitable kidney donors. In Chichigalpa, however, heat is rending these socio-organic ties asunder, even as the disease that affects Jorge and Ulises spreads around the world, linking them to Sri Lankan and Egyptian farmers (R. Chatterjee 2016). In the case of CKDnt, heat is not so much a cause (an external factor acting on bodies and systems) as a catalyst (a mediating force that turns potential relations into actual ones). People at work are constantly exposed to heat, but heat also exposes them to the forms of solidarity that make and unmake them, from kinship ties to paternalistic plantation structures to caloric loads to the crystalline form of refined sugarcane. In making this claim about heat, I am adapting Marilyn Strathern's ideas about eating in Melanesia: "Eating does not necessarily imply nurture; it is not an intrinsically beneficiary act, as it is taken to be in the Western commodity view that regards the self as thereby perpetuating its own existence. Rather, eating exposes the Melanesian person to all the hazards of the relationships of which he/she is composed" (1988, 294).

Strathern's view of eating as exposure leads us back again to the stories of those other, perhaps less visible workers—the unwaged and the self-employed like doña Julia and migrant workers like her younger kin. These workers are not simply suffering in silence. They are working with heat, too. They are looking, collectively, for ways to endure its compounding effects on kidneys and relationships, even if they do not have a unified political agenda or a movement behind them. Doña Julia's children are called home by the heat of the hearth from the heat of dangerous low-wage jobs. The making of a life beyond work depends on heat, too. The challenge for ethnography is to link intimate and transnational forms of solidarity, without privileging that which becomes familiar to us as "labor activism" over that which seems to be familiar as quiet "social suffering."

This predicament is emblematic of what Kim Fortun (2012) calls "late industrialism." In late industrialism, heat is both evenly distributed and patchy. Some bodies—not all—are becoming exhausted, yet in a discourse that runs from the Industrial Revolution to contemporary anxieties about the Anthropocene, we are supposed to imagine an entire biosphere in peril. What nineteenth-century philosophers called the "heat death of the universe"—the ultimate victory of an unruly nature over the efforts of human work—seems like a possibility (Rabinbach 1992).

Where does anthropology stand here? An approach that simply follows caloric energy through circuits of production and consumption elides heat's role as an index of vital energy—the kind of energy that, as Nicaraguan cane workers used to say, outlived the human body and reanimated cattle (Gould 1990). A perhaps more critical approach might lead us to identify heat not as "natural" but as a "social determinant" of health. That approach, however, risks positioning the heat exposure of plantation workers like Jorge and Ulises, self-employed workers like doña Julia, and migrant farm and construction workers in the United States and elsewhere as only "suffering" (Holmes 2013; Farmer 1999). What gets missed in both these approaches are the subtle ways in which heat is not only a cause of suffering but also a catalyst for making suffering socially knowable and thus actionable. There is no numerical degree marker at which solidarity breaks down, just as there is no such point at which the bonds of kin, class, or politics are forged.

In physics, heat signifies waste, but it also signifies resistance (White 1995b). Turning heat into a problem of technical management, thermodynamic control, or simply unequal exposure risks reducing people like doña Julia, Jorge, and Ulises to their status as workers. Seeing heat as that which enables us to think relationally—as that which allows us to see our social connections before they are objectified in objects or actions—is a way of pushing back against this reductionism.

Seeing heat in this way also helps to "provincialize" it (Morrison 2015). Is the heat that makes Salvadoran cane cutters sick the same heat that makes

Nicaraguan cane cutters sick? Is the heat that causes coral reef die-offs the same as the heat that causes CKDnt, or the heat that causes the bodies of Central American and Mexican labor migrants to decompose after they perish in the Sonoran desert (A. Moore 2016; De León 2015)?

I think doña Julia would agree that they are not the same. When I come to her house, she says, "Que calor." Posed as a question, "Que calor?" means "What heat?" Posed as a statement, it translates best as "this heat."

This heat. The stories of heat from Nicaragua and beyond might lead us to follow feminist scholars in imagining nature's work and human biology in a more situated manner (Lock 2017). A cue from Karen Barad's (2007) reading of physics and performativity in an age dominated by quantum mechanics, rather than the Newtonian mechanics that led to the physical association of heat with work in the industrial age (Rabinbach 1992), suggests that CKD and CKDnt are not mere "effects" of uneven labor processes in which heat is a wasting agent; they are also the "causes" of new understandings of heat in, of, and as work. Likewise, heat, whether as the waste emitted by inefficient machinery or as the sunlight that photosynthesizes sugarcane, is not simply a cause of suffering. It is also an effect of labor, activism, epidemiological studies, and corporate and state resistance to all of the above.

Ethnographic engagement that refuses a clear distinction between cause and effect suggests that heat's power to shape life forms and forms of life is contingent and differential. A consideration of the lives and deaths of *these workers* rather than workers in general, offers insight into the massively distributed yet patchy effects of late industrialism. Heat both reflects long-standing forms of discrimination and difference and produces new forms of solidarity. Ethnography, then, can break down the metaphorical and social forms of insulation that make the work of heat seem sometimes global, sometimes invisible. It can help us find patterns and points of pushback within what look like random flare-ups, combustions, and burns.

CHAPTER SIX

Metabolic Relations

Korean Red Ginseng and the Ecologies of Modern Life

ELEANA KIM

It was December 2011, and I was riding with Mr. Lee, a researcher with DMZ EcoResearch, a small nongovernmental organization (NGO) composed of citizen ecologists who monitor the flora and fauna in the western regions of the demilitarized zone (DMZ) in South Korea. DMZ EcoResearch has permission to regularly enter the Civilian Control Zone (CCZ), a militarily restricted area immediately adjacent to the DMZ where training grounds, minefields, and firing ranges exist side by side with agricultural tracts. That day, our task was to count endangered cranes, which use the Han River estuary as a wintering site in their annual migrations. In particular, we were looking for red-crowned (*durumi*) and white-naped cranes (*jae durumi*). The red-crowned cranes breed in northeastern Russia and arrive in the DMZ in mid-October, staying until early March, and the white-naped cranes use the DMZ as a stopover on their way to their wintering sites in Japan. These endangered cranes winter in the DMZ and CCZ (together referred to as the DMZ area) where they have ready access to food—marine and freshwater creatures in the shallow streams and wetlands of the estuary and rice grains left over from the autumn harvest.

Monitoring the cranes with DMZ EcoResearch meant driving along the backroads and scanning across the tawny landscape of reaped and threshed rice stalks, looking out for the birds' distinctively graceful forms—long legs and small heads—oftentimes bent over and gleaning grains in the frozen ground. Along our route, ginseng fields—large patches of blue or black tarp fragmenting the landscape—would invariably provoke laments from my interlocutors who condemned their ecologically deleterious effects. Ginseng must be shade grown, and the tarps are attached to short wooden frames to block the sun. For my ecologist friends, these fields were not just aesthetic blights on the landscape but were direct threats to the survival of the cranes, as the shade-grown crops

reduce the amount of land the birds can use to rest and feed, and ginseng growers introduce harmful chemical fertilizers and pesticides into the soil. Blue-
and black-covered landscapes thus represented for them the most egregious
examples of human disregard for the lives of nonhuman species.

Land speculation has been continuous since the late 1990s liberalization of
the CCZ, in which "outsiders" (*oeji in*), as they were referred to by my friends,
purchase plots in anticipation of a unified peninsula, projecting a future in
which these areas will become a new frontier of development and economic
growth. Until that time, landowners seek productive use of the land and have
been increasingly planting ginseng, a crop that, unlike rice, is consistently high
profit and relatively low maintenance. In fact, around this time, "DMZ Ginseng"
began appearing on the market, foregrounding an array of old and new symbolic registers—including militarization, pristine nature, and the terroir of the
DMZ region, which is in close proximity to Gaeseong, the ancient capital of
ginseng production (now on the North Korean side of the DMZ).

In the past two decades, formerly off-limits areas have opened to agricultural production and tourism development. Drawing on associations with war
and division but also linking tourism to agriculture—tour companies now offer
"farm stay" packages alongside conventional "security tourism" of the Joint
Security Area—DMZ consumables have proliferated. Ginseng is simply one
in a growing list of food products branded as "DMZ," including honey, rice,
apples, beans, and even springwater. The contemporary DMZ's associations
with pure and clean nature have become so naturalized that the previously dark
image of exploded landscapes, human remains, and military pollution seems to
have entirely faded to the background. This newly foregrounded "pure" DMZ,
however, masks the ways that militarized ecologies in fact support the proliferation of capitalist value production.

My conversation with Mr. Lee that day had started predictably enough—
we commented on the ginseng fields, their seeming ubiquity, and the dangers
they posed to the birds. The highest quality ginseng is grown for six years, and,
unlike in other parts of Korea where ginseng is cultivated, growers in the CCZ
do not have to worry about thieves, as the area's military securitization not
only protects the nation from attack from North Korea but also protects this
highly profitable crop from pilferers. I learned a few other facts—the saponins
in ginseng (ginsenosides) are restorative for human bodies but are also chemical compounds that repel consumption by other animals due to their bitter-
ness and sometimes fatal toxicity. And ginseng growers move their fields after

harvesting, leaving behind depleted soil to colonize new habitats that might otherwise be used by the endangered cranes. Thus, the proliferation of ginseng fields essentially ensured that birds would lose sources of food and resting areas for years to come. Ginseng was also making inroads into areas east of the Han River estuary, into the more mountainous areas of Yeoncheon. My colleagues had been speculating that the cranes, which were increasingly hard to find around the estuary, were moving to Yeoncheon in order to locate more hospitable wintering sites, and now they worried that the ginseng fields would put pressure on their habitats there as well.

Having heard all this dire information about the negative ecological effects of ginseng, I had to ask: did Mr. Lee himself use ginseng? I was, in a strange way, surprised by my own lack of surprise when he answered in the affirmative. There is something quintessential about the consumption of ginseng among South Korean men, and South Koreans as a whole constitute the world's largest domestic market for ginseng, totaling over US$1 billion per year. In fact, my ecologist friends reflected commonly held nationalist sentiments when they complained that Chinese ginseng and counterfeits were flooding the Korean market, as well as diluting South Korea's global market share. And one could also connect the ginseng production in the CCZ that they deplored to the high profitability of ginseng and its expansion into new markets across Southeast Asia, Europe, and North America.

Metabolisms: Naturecultures and (Human) Bodies

In its blind and measureless drive, its insatiable appetite for surplus labour, capital oversteps not only the moral but even the merely physical limits of the working day. It usurps the time for growth, development and healthy maintenance of the body. It steals the time required for the consumption of fresh air and sunlight. It haggles over meal-times, where possible incorporating them into the production process itself, so that food is added to the worker as to a mere means of production, as coal is supplied to the boiler, and grease and oil to the machinery.
— Karl Marx, *Capital, Volume 1*

Marx famously wrote, "Labour is, first of all, a process between man and nature, a process by which man . . . regulates and controls the metabolism between himself and nature" (1976, 283). He furthermore linked capitalist production with the growth of cities and charged it with "disturb[ing] the metabolic interaction between man and earth, i.e., it prevents the return to the soil of its constituent elements consumed by man in the form of food and clothing; hence it hinders the operation of the eternal natural condition for the lasting fertility of the soil" (637). In doing so, he identified a vicious cycle in which soil exhaustion required industrial solutions, which "undermin[ed] the original source of all wealth—the soil and the worker" (638). This notion of "metabolic rift" has been expanded and applied since the late 1990s to the ongoing ecological crisis by John Bellamy Foster, Brett Clark, and Richard York (2010), who extend the metabolic rift to an "ecological rift" that encompasses the various crises associated with the Anthropocene. For Marx and ecological Marxists like Foster, Clark, and York, metabolic rift connects capitalist forms of extraction and accumulation with ecological collapse.

The cultivation and processing of ginseng in the Korean peninsula have, since the Japanese colonial period, followed a plantation and industrial factory model. Ecologically unsustainable and destructive of the local ecologies, ginseng fields seem to exemplify Marx's "metabolic rift," especially given the way that ginseng cultivation leads directly to soil exhaustion. Ginseng might therefore appear to be a perfect example of this declensionist narrative, especially given the crop's deleterious effects on soil quality and the alienated, unsustainable plantation model that structures its production. Indeed, the image of tiny human-shaped roots colonizing the soil and depleting it of its nutrients to be then consumed by overstressed consumers in the overcrowded megacities of South Korea could hardly be a more fitting illustration of metabolic rift. Once it is processed into *hongsam*, or red ginseng, its efficacy as a panacea for stress-related maladies makes it an apt remedy for the alienation and exhaustion produced by capitalism, functioning like "coal is supplied to the boiler and grease and oil to the machinery" (Marx 1976, 376). Yet rather than simply mapping a closed system in which ginseng exemplifies the paradigm of metabolic rift within a capitalist teleology, it is worth considering how ginseng works—as a naturalcultural assemblage within complex ecologies, metabolic relations, and multispecies worlds.

For Hannah Arendt, Marx's valorization of labor as the basis of man's "species-being" restricted human existence to that of the "private." For her, Marx's equation of labor with "the metabolism of man and nature" meant that it was "not freedom but compulsion . . . that makes man human" (Arendt 2002, 286). She writes, "People who do nothing but cater to these elementary coercive needs were traditionally deemed unfree by definition—that is, they were considered unready to exercise the functions of free citizens" (285). For Arendt, freedom from labor and biological needs was a prerequisite to her vision of work, "the activity which corresponds to the unnaturalness of human existence, which is not embedded in, and whose mortality is not compensated by, the species' ever-recurring life-cycle" (285). As others have noted, Arendt's politics is predicated on human exceptionalism and a stark binary between "nature" and "culture" and "public" and "private" (see also Hartigan, this volume).

In a post-foundational moment, in which the nature/culture binary has been undone, Arendt's firm distinction between work and labor likewise dissolves, and her notion of "work" and politics might then be extended into the realm of naturecultures and the more-than-human. This approach resonates with Jason W. Moore's (2016) vision of the Capitalocene, which draws on Donna Haraway (2016) and others to argue that capitalism itself is "world-ecological" and "a multispecies affair." With this in mind, we can approach the metabolic relations of Korean ginseng as "working" across multiple scales and entanglements, from the microbes and fungi that cohabit the soil and feed on ginseng roots, to human bodies laboring to cultivate ginseng, to the birds and other creatures displaced or disappearing along with the rice paddies, to the marketing of ginseng's miraculous qualities to postindustrial consumers, to the chemical compounds metabolized and made bioavailable (or not) in human digestive tracts.

Korean ginseng's metabolic relations of human labor, nonhuman work, and capital are increasingly tied to the militarized landscapes in which it is grown, and what I call a process of terroirification, in which the securitization of space lays the ground for the appropriation of value from the ecological relations that constitute the "DMZ's nature." The CCZ houses the majority of the 600,000 South Korean and 28,000 US troops, and soldiers serving on US military bases around the world regularly converge for training exercises in the world's largest military training grounds and firing ranges that are located in this area. Thus, the "pristine" DMZ and CCZ are also the most heavily contaminated with military waste—including more than one million land mines, countless unexploded ordnance, and explosive remnants of war—a full accounting of which will most likely never be made. The DMZ, one of the most heavily militarized buffer areas in the world and a relic of the Cold War, once excerpted "nature" from capitalist exploitation while protecting South Korea's capitalist system from Communist aggression. Increasingly, however, its value as an "ironic" preservation zone of rare biodiversity is being converted into "cheap nature," which itself depends on the DMZ's securitization of space (J. Moore 2016).

Terroir is most often associated with inalienable connections between place and taste, originating in French viniculture. As Heather Paxson writes of artisanal cheese producers in Vermont, terroir offers a "theory of how people and place, cultural tradition and landscape ecology, are mutually constituted over time" (2010, 444; see also Trubek, Guy, and Bowen 2010). Although it is not a culturally salient term in South Korea, terroir is a useful framework that helps me to describe the processes of converting human-nonhuman metabolic relations into the "work of nature," as a form of value production. In the case of ginseng, which is a highly valued commodity, holistic notions of landscape—including soil, climate, topography, place, identity, and tradition—are being economically and ecologically amplified through their proximity to the militarized border.

Cultural narratives of the (South) Koreanness of Korean ginseng tie terroir to techne in ways that resonate with indigenized forms of agricultural nativism and organicism expressed in the Korean aphorism of *sint'oburi*, which literally is "(human) body, earth, not two." However, in the highly scientized representations of ginseng, the circulations of which are largely controlled by a single monopoly, the harmonious relationship of farmer and land, which is central to sint'oburi discourses, is largely absent. Indeed, human labor is displaced by the labor of the root itself, the ideal form of which is, ironically, a human body. Following a discussion of the history and cultural significance of ginseng, I argue that the work of ginseng is being organized by local and national governments as well as corporations and consumers to terroirize the DMZ area, converting a multispecies ecology, characterized by militarized security and restrictions on human activity, into a space of South Korean capitalist expansion.

What Makes Ginseng Korean

The export of Korean ginseng dates to premodern times, with historical records and contemporary lore claiming Koreans' consumption of ginseng for more than two thousand years. Ginseng was presented as gifts to Ming China as well as Tokugawa Japan during the Joseon dynasty (1392–1897). Soyoung Suh succinctly sums up ginseng's "commercial and cultural significance" by stating, "No other herbs match ginseng in conveying the quintessence of Korea" (2008, 397). Wild ginseng grows in well-drained soils in shady, deciduous hardwood forests and is notorious for being difficult to cultivate. An American ginseng craze in the late 1800s led to many disastrous failures, as farmers attempted to cash in on the Chinese appetite for North American ginseng (Johannsen 2006, chap. 1; Carlson 1986). In Korea, the mass cultivation of ginseng became standardized and regularized under the Japanese colonial government, which established ChungKwanJang ("government certified") ginseng in 1899.

The historic capital of ginseng production was Gaeseong, which became part of North Korea after the Korean War. Part of the official historical narrative as told by the Korean Ginseng Corporation (KGC) includes the heroic efforts of three South Koreans who risked their very lives to bring back ginseng seeds to the South in February 1952. These seeds ensured that superior ginseng production would be able to continue in the South, with what is now the KGC's factory, established in Buyeo in 1956. By the 1960s, ginseng was being exported to Hong Kong, Thailand, and Taiwan, and soon to other Asian countries as well as Europe and North America.

Ginseng (*insam*) is a plant whose roots have been recognized in traditional Asian medicine as a cure-all since at least the end of the Han dynasty (206 BCE–270 CE) (Kuriyama 2017). Among the four main species of ginseng, only one is grown in Korea, *Panax ginseng* C. A. Meyer. What makes Korean ginseng distinctive, according to experts like Kim Si-Kwan, professor of biomedicine and former president of the Korean Society of Ginseng, is both the particular characteristics of the Korean peninsula's climate and soil and the cultivation techniques that have been perfected over a millennium. These techniques, which involve the transplantation of year-old saplings at a forty-five-degree angle and their growth for at least six years, have resulted in the distinctive "person"-like shape of Korean ginseng, with a robust "body" and two strong "legs," with smaller roots extending out from those main body parts. This particular shape is also related to an abundance of chemical compounds that make ginseng medicinally efficacious. If *Panax ginseng* C. A. Meyer becomes "Korean" through human know-how, or techne, then it becomes "Korean *red* ginseng," or hongsam, through the particular processing technologies performed by women workers in the largest ginseng factory in the world, using technoscientific interventions that have been proven to amplify the effectiveness of its chemical compounds. Six-year-old ginseng roots are steamed and then dried, producing a ruddy, reddish color that is also identified with a high concentration of ginsenosides, which have a steroid-like effect that can counteract fatigue and stress.

In the past decade, ginseng's medical uses and market value have been linked to an explosion in scientific studies that attempt to isolate and test ginsenosides at the cellular and molecular levels. The turn from identifying ginseng saponins with increased energy to seeking their effectiveness in regulating specific biological processes reflects what Hannah Landecker (2013) identifies as a shift in the cultural logic of metabolism from industrial to postindustrial. In the thousands of studies published annually, saponins are linked to conditions as diverse as hypertension, stomach cancer, menopause, sexual dysfunction, dementia, swine flu, MERS, and diabetes. An ever-widening set of applications for ginseng and the overall expansion of the global herbaceutical market have contributed to the exponential growth in hongsam sales in South Korea and in markets around the world. Research scientists like Kim Si-Kwan publish their findings in medical journals and present their findings at the annual International Symposium on Ginseng, as well as in the open access *Journal of Ginseng Research*. Kim also helps to promote ChungKwanJang, produced by the KGC, in places like Singapore, one of the forty countries to which South Korean red ginseng is being distributed and marketed.

The KGC holds a pride of place in Korean red ginseng history. Once part of the state-run Korean Tobacco and Ginseng Corporation, which had roots in the Japanese colonial government, KGC was spun off from the KT&G in 1988. The state monopoly was dismantled in 1996. Yet its modernization and rationalization under the Japanese still inform its brand identity and market value, as concerns about the quality and authenticity of Korean ginseng that existed at the turn of the century continue to plague the ginseng market. Thus, like the "green milk" in China that Megan Tracy (2013) describes, Korean red ginseng has long embraced both terroir and technoscientific production, without viewing one as canceling out the other. The role of the state, in the form of the Japanese colonial government then and the South Korean state after the national division, has been to act as a guarantor of both the technoscience and the origins of Korean red ginseng and to protect against counterfeits. ChungKwanJang (CKJ), invented under the Japanese, is the brand name that retains this state-science-terroir nexus of authenticity. Meaning "authentic government certification" (translated as "health code" by KGC), it served as a government- and now market-based seal of approval. The effectiveness of Korean ginseng is framed by the KGC as a conjuncture of terroir, ancient cultivation techniques, and modern processing technologies.

Today, CKJ is the brand that keeps KGC in the top position in the global market, though it now has competition from other producers and distributers within South Korea, most notably Nonghyup, the country's largest farming association. The ginseng market grew rapidly in the 2010s with 2,978,000 tons of ginseng produced in 2014, representing a four-fold increase from 2011. Nevertheless, a sense of anxiety and crisis characterizes the market, with production and demand waxing and waning with fickle consumer tastes and health and well-being fads. Nevertheless, recent surveys show that more than 40 percent of South Koreans between the ages of twenty and fifty take dietary supplements, and red ginseng is among the most popular products, alongside multivitamins and omega-3s. Given the unpredictability of the domestic market, KGC and other companies have deepened their focus on the global market as competition from other countries, particularly China, intensifies.

One effect of this competition has been the proliferation of products and niche markets—ginseng can now be consumed through a myriad of delivery vehicles. Traditional forms like the dark, unctuous elixirs or the actual steamed and dried roots and freeze-dried crystal powder teas are now joined by candies, crackers, cosmetics, beverages, lozenges, gummies, capsules, and powders. The appearance of DMZ-branded ginseng is the latest stage of the liberalization of Korean red ginseng in a context of heightened global competition.

Labor and Work of Ginseng

The cultural associations of hongsam in South Korea are overwhelmingly connected to the regulation of the body, with the primary targets the immune system, cognitive functions, and energy/fatigue. In the United States, where it is often sold near the cash registers at drugstores and gas stations, it has long been identified with boosting male virility, but in South Korea, its energetic, immunological, and metabolic effects are seen as beneficial for all. It is a frequently gifted commodity that is especially promoted during the Korean Thanksgiving (*chuseok*) and Lunar New Year. KGC markets products to parents for toddlers, tweens, and teenagers, including a supplement called "I-Pass" that advertises "enhanced memory functions" for students stressed out about schoolwork and exams. Products target every stage of the life cycle, from children as young as three years old through high school, then to stressed-out college students, office workers, and elderly people suffering from fatigue or fading memory. In 2016, KGC even inaugurated a line of dog kibble, Ginipet.

In traditional Korean medicine, ginseng is considered to be a "superior" medicine, meaning it can be consumed by almost anyone without adverse effects. Whole dried roots of CKJ can be as much as $250, and packages of elixirs and concentrates are upward of $65. The daily dose packages encourage the quick consumption of ginseng, and these are marketed for a population that is continually on the go, without time to stop and dissolve a packet of ginseng tea or boil a root for a more traditional preparation.

The abolishment of the state monopoly in 1996 has led to a diversification of the market, including farms that promote "organic" ginseng and have online stores with direct-to-consumer marketing. Yet the CKJ brand is still the most recognizable and the most trusted (its origins in the Japanese colonial period notwithstanding) among consumers, especially as more and more brands appear, claiming the same chemical compounds but lacking the long legacy. CKJ also requires its farmers to comply with regulations regarding pesticide residue, unlike other manufacturers. Images of ginseng fields and the plants themselves, which have bright red seeds in their third year of growth and sprout three-leaf stems every year, are common, but the shaded fields are presented as the main spectacle, rather than the farmer who might be tending them. Agency is located in Korean terroir itself; that is, the ideal soil and climate for ginseng cultivation and the more than one hundred years of know-how that CKJ represents.

There is a well-developed discourse and romantic imagery of farmers in South Korea connected to sint'oburi, so the absence of farmers in the discourses and imagery of the ginseng industry is notable. Instead of happy faces of smiling farmers presenting the fruits of their labor, wide stretches of ginseng plantations are the main protagonists in advertisements like the one produced in 2014 featuring soap opera actor Ahn Seong-gi.[1] The sepia-toned ad shows him, alone, in a seemingly boundless ginseng field, depicted from an aerial view that shows a sweeping expanse of the characteristic shade-covered land. Each black square of shade, angled against the sky, constitutes a pixel in a larger picture of the ginseng plantation.

Ahn is alone in this serene monoculture, with one beetle and a single frog as the only other creatures in this surreal landscape. His smooth yet slightly husky voiceover suggests an attitude of awe, or even sublimity: "Hongsam isn't grown simply from human effort. The power of the sun, rain, and wind are all called on. For well over one year, the power of the earth is nurtured, attaining the requirements for growing ChungKwanJang hongsam." From "nature" to the techniques of KGC and its 260 tests, run seven times, on all of its ginseng plants, we learn of the quality control that is required to earn the brand CKJ. But we are also reminded that even after a hundred years of making hongsam, not all of its secrets (*bimil*) are known: "It's not due to chance or out of greed, but from the best efforts of people. There is no other way than that. When it comes to hongsam, there are no short cuts. CKJ takes the proper path." As Sarah Besky notes in this volume, "the work of monoculture is maintenance," and in this ad, monoculture's homogeneity has been elevated to an aesthetic of human aspiration for perfection.

In this and other CKJ narratives, "nature," particularly Korea's climate and earth, is the basis for hongsam's quality and effectiveness. Despite the obvious human intervention of the shade-grown ginseng farms, human labor is conspicuously absent, and when humans appear, they are either consumers or scientists in white lab coats describing the effects that ginseng has had on laboratory animals (most commonly mice) or in human clinical trials, which have proven the effectiveness of Korean red ginseng for a wide variety of conditions.

Thus, modernist industrial methods notwithstanding, the corporate story of ginseng is a peculiarly hands-off version of human mastery over nature, one of nurturance and patience, scientific testing, and quality control. But this is still a highly alienated landscape, in which the actual human labor that produces the plantations is excluded, creating a fetish of the plantations themselves. When farmers do appear, they are presented as biopolitical subjects who must be disciplined to conform to KGC's stringent requirements. Indeed, it is not the ginseng that is the main object of the 260 tests performed by KGC but the farmers whose crops can be rejected or devalued if they fall short. In these corporate representations, they do not represent embodied expertise through authentic connection to the land. Instead, they appear as precarious workers who are at the mercy of ginseng's difficulty as a crop and who are "frightened" by the KGC inspectors and their high standards.

Given the centrality of soil or earth (*ddang*) to the terroir of Korean red ginseng, the unsustainable metabolisms of ginseng in its earthly environment are worth noting. Ginseng is a highly intensive root that strips the soil of all nitrogen, so that by the time it is harvested, the soil has been depleted. Two to three years of crop rotation are necessary to replenish the soil nutrients, and two years of soil disinfection and fumigation are required to prevent blight. Blight presents the greatest risk to farmers because of the pathogen *Cylindrocarpon destructans*, a fungus that lingers in the stray roots and detritus from the previous year's crop and causes root rot (*geunbu byeong* or *insam bburi sseokeum byeongwon'gyun*). The disease is also known in English as "disappearing root rot" because, at its most robust, *Cylindrocarpon destructans* eats out the root, leaving just the skin behind. The chlamydospores of this fungus can live for more than ten years, meaning that the probability of replant failure is very high. Even the Latin name of this fungus suggests a hierarchy of value that privileges the economically valuable life of the ginseng root, and relegates *Cylindrocarpon destructans*'s labor to that of mere destruction; a destruction, one might add, that cannot be easily assimilated into the creative destruction of capital.

Root rot, combined with soil nutrient depletion and climate change, means that growers are increasingly heading north in search of new cultivation areas. Moreover, with declines in the price of rice, due to weakening demand and cheaper imported rice, farmers have been switching to other crops, with ginseng, given its high profit margins, among the top alternatives. Scientific research has also suggested that rice-paddy-converted ginseng fields yield high-quality ginseng and reduce the threat of *Cylindrocarpon destructans*. For a nation marked by overdevelopment and population density, few areas could provide better "virgin soil" than the militarized borderland of the DMZ.

Terroirification of Ginseng

The Paju Gaeseong Ginseng Festival was inaugurated in 2004 and now takes place annually in mid-October, during the harvest season. Paju is a city located on the western side of the peninsula, just south of the DMZ, and the municipal government has been active in capitalizing on this proximity to boost its tourism revenue for several years. The festival is a key event that brings thousands of visitors to Imjingak Park, which is the northernmost location that domestic tourists may travel to and is a peace park that commemorates the seven million families separated by the Korean War. At this site, a complex politics of division, tradition, and terroir are on display, with ginseng-themed cartoon characters, K-pop concerts, traditional Korean wrestling competitions, and the enactment of a historical drama depicting the life of Heo Jun, the seventeenth-century author of *Dongui bogam*, a compendium of Korean traditional medicine that is considered to be a major cultural and scientific achievement and was recently registered by UNESCO.

Rather than the reverse engineering of terroir that Paxson (2010) describes among artisanal cheese producers in Vermont (who select Old World cheese varieties that will match the given characteristics of their farmland), in this case, the connections between land, culture, and people that constitute terroir are constructed through a set of historical elisions and equivalences, as exemplified by the festival's promotional language:

> Paju Gaeseong Ginseng Festival: DMZ . . . where Gaeseong ginseng lives and breathes! Six-year-old Paju Gaeseong ginseng is grown in the clean and pure areas of the Civilian Control Zone and Gamak Mountain. . . . Six-year-old Paju Gaeseong ginseng is the representative ginseng of the Republic of Korea, and continues the legacy of Goryeo ginseng. Come see for yourself the top-rated Paju Gaeseong ginseng and Paju City's guaranteed best quality six-year-old ginseng.[2]

Paju's status as a border area near the DMZ creates a geographic connection to Gaeseong, North Korea, the ancient capital of ginseng production in the Goryeo period (918–1392 CE), where ginseng continues to be grown and processed into hongsam, an important export commodity for North Korea. Claiming spatial contiguity and historical continuity, this narrative collapses cultural and political distance to naturalize Paju (South Korean) ginseng as authentic Gaeseong (pre-division Korean) ginseng. In the process, Paju ginseng seeks to combine the ancient terroir of Gaeseong with the contemporary branding of the DMZ and its associations with "pure nature," suggesting that Paju/Gaeseong's "ecofriendly" ginseng is perhaps even better than the real thing—more modern and healthy than North Korean Gaeseong Goryeo ginseng. As I have suggested, a key element in the terroirification of ginseng is the absence of people (growers or pickers) and, in the case of the DMZ, soldiers—who greatly outnumber the population of civilians living in just two villages on the western side of the CCZ. In this way, the value of DMZ ginseng relies not only on the symbolic naturalization of landscape, place, and identity but also on the elision of the division and the erasure of militarized ecologies, which are its very conditions of possibility.

Conclusions

In the context of South Korean hypercapitalism—its culture of relentless innovation and overwork combined with an intensified sense of physical fragility and immunological risk—ginseng can be seen as the supplement du jour. It is this connection between South Korea's hypermodernity, global capital, and ginseng's "adaptogenic" efficacies that became clear to me in my conversation with Mr. Lee, who was in his early fifties when we met. He was one of the founding members of DMZ EcoResearch, having joined in 2005. He was at that time a restaurant owner, with two successful kimchee stew restaurants, even though he had a master's degree in marine biology and worked for a government science research institute in his late twenties. After serving in the military, he went on to own a livestock factory and then opened his restaurants in a town close to the CCZ. In the course of his personal narrative, he emphasized how unattractive the highly competitive society, concentrated in the megalopolis of Seoul, was to him. He, like others in the group, had experienced health problems because of work-related stress, and his connection with the NGO was part of his own attempt to remedy his ailments through a connection with nature. He explained how, unlike most parents, he didn't send his children to cram schools (*hagwon*), instead letting them discover their own interests.

Mr. Lee's narrative resonated with another member of the NGO, Mr. Kim, who was fifty years old and often talked about how his experiences in nature had saved him at a crucial moment in his life. He wanted to bring more city people to experience the rare ecology of the DMZ because they were so out of touch with nature. For him, nature's restorative qualities were a remedy for the stresses of modern life. He also consumed traditional Korean medicine even though he never failed to rail against the ginseng fields and their detrimental ecological effects. Like Mr. Lee, Mr. Kim's personal narrative was marked by accounts of how he tried to avoid the hypercompetitive toxicity of South Korean society.

The purpose of highlighting this seeming discrepancy is not to suggest that Mr. Kim and Mr. Lee are hypocritical, as there is no need to expect that consumption behaviors and ideological beliefs necessarily align. Instead, I'm interested in how the ecology of ginseng cultivation, and its long histories of terroir as both place making and culture making, are connected to broader economic subjectivities and relations of capital. Mr. Lee's and Mr. Kim's relations to ginseng remind me of Jacques Derrida's reading of Plato's pharmakon, in which "operating through seduction, the pharmakon makes one stray from one's general, natural, habitual paths and laws" (1981, 70). For Derrida, the pharmakon is at once poison and remedy, the illness and the cure, and the mark of ambivalence in all writing.

In this chapter, I trace the multiple and heterogeneous relations among people, places, nonhuman species, economies, affects, and bodies that are at once caused and cured by ginseng as a heavily marketed panacea for the stress-induced conditions of contemporary life. Looking for a respite from South Korea's compressed modernity, people like Mr. Lee and Mr. Kim find it in the already ambiguous sites of "nature" in the DMZ, the most heavily fortified border in the world. Seeking rare species like the migratory cranes, they deplore the destruction of their habitats as it is most immediately and visually represented by the proliferating fields of ginseng. Ecologically, the terroirification of DMZ ginseng is destroying the symbiotic relations between cranes and humans, materially fragmenting ecosystems, including the forests where wild ginseng might grow. Here, the model of industrial farming brings with it an ethic of alienation that includes human labor and the privatization of land, as well as the destruction of the very terroir that is a central aspect to the KGC brand.

Moreover, the monocultures of ginseng, and the individuated pods required to produce the perfect "human form" valued by cultivators, spatially and ecologically alienate nonhuman species, most notably cranes, which are rapidly disappearing from their historic wintering sites. Mr. Lee and Mr. Kim blame ginseng for the attenuation of nonhuman dependencies and relations, a moral discourse in which connections to the organic world are the remedy for the ills of life and labor in hypercapitalist South Korea, even as they consume ginseng on a daily basis. Indeed, as KGC promotes the consumption of ginseng in developing nations, one can only imagine that what may be most "Korean" about Korean red ginseng is its associations with South Korea's advanced capitalist society and its citizens' intensified attempts to "adapt" their bodies and minds to the market-based anxieties inherent to capitalist economies of overconsumption.

Korean red ginseng is a multispecies assemblage that emerges out of moral, affective, material, and economic relations and configurations from the fields of speculation along the CCZ to the consumption of ginseng for people like Mr. Lee and the global branding practices of the KGC. The terroirification of Korean ginseng in the DMZ is part of a wider set of practices that attempt to purify Korean ginseng, fix its qualities, and "retro-botanize" it (Mukharji 2014). As Projit Bihari Mukharji writes, "The variety of pasts and the different identities they are entangled with inspire new futures of use" (2014, 80). Terroirizing ginseng in the DMZ and its ancient techne link its present to Gaeseong and its future value to developing nations, in which ginseng's chemical "adaptogens" or stress-relieving efficacies may, in fact, reveal it to be like Derrida's (1981) pharmakon—both poison and remedy. Ginseng's particular metabolic relations, which are part and parcel of a risk society under climate-induced duress, complicate the ecological imaginaries of Mr. Lee and his colleagues, for whom nonhuman nature and its conservation, like that of the birds' wintering habitats, are the very remedies they need to get by in South Korea.

Notes

1. The advertisement is available on the Korean Ginseng Corporation's YouTube channel (https://www.youtube.com/watch?v=ZPHTk1S5GXY). Accessed December 30, 2018; translation of quoted text by the author.
2. Paju Gaeseong Ginseng Festival website (https://tour.paju.go.kr/tour/festival/festival02/festival02_1/festival_02_1_tab1_1/festival_02_1_tab1_1_01.jsp). Accessed December 30, 2018; translation by author.

CHAPTER SEVEN

How Guinea Pigs Work
Figurations and Gastro-Politics in Peru

MARÍA ELENA GARCÍA

The guinea pig will save Peru.
—Lilia Chauca, director of the Guinea Pig Project,
National Institute in Agrarian Innovation

Guinea Pig Power

Lilia Chauca, director of the Guinea Pig Project at the National Institute in Agrarian Innovation (INIA), wears her expertise lightly as she describes the state of the Peruvian guinea pig (*cuy*) in her office in Lima. She is proud of the work she and her team have done in the "genetic improvement" of guinea pigs and in the development and intensification of cuy production. She established the Guinea Pig Project in 1968 with her husband, the late Marco Zaldívar, and with support from the US Agency for International Development and the University of North Carolina (Morales 1995).[1] The project focused on teaching Indigenous peoples how to better manage their guinea pigs—how to pick which ones to eat and how to be more intentional and efficient about breeding. They also began developing specific breeds that would grow larger and more quickly and females that could give birth to higher numbers of young in order to maximize productivity. Today, the INIA and the Guinea Pig Project headed by Chauca are widely recognized as the most important sites for guinea pig breeding and production in the country. But as Chauca told me, it is only recently, in the context of the so-called gastronomic revolution that has enveloped Peru, that the power of the guinea pig has become visible. "Now is our moment," she told me in 2012. "The guinea pig will save Peru."[2]

The moment Chauca mentions is one of national hope and possibility.

131

Peruvian food—fusion cuisine in particular—has arguably become the central element in Peru's national project. Narratives disseminated nationally and
internationally via print media, film, culinary tours, restaurant websites, tasting
menus, and cookbooks tell a story of food uniting Peruvians, erasing centuries
of racial antagonisms, and resituating Peru politically, economically, and culturally. Peruvians have embraced this moment of culinary fusion as a way to
consume, perform, and celebrate sovereignty in the wake of decades of political violence that threatened the very idea of a national project and community.
The war between revolutionary forces and the state, lasting officially between
1980 and 2000, led to at least seventy thousand deaths and countless unjustly
detained, tortured, disappeared, and displaced people. Three-fourths of those
impacted were Indigenous peoples (CVR 2004). This is crucial background to
contemporary stories of economic and cultural success. Historically, the racialized geography of Peru was such that elites in coastal Lima understood (and
cared) little about the horrors Indigenous Peruvians had experienced. Over the
past fifteen years, Peruvian elites and state actors have embarked on a national
project that attempts to reclaim Peruvian sovereignty and re-narrate this recent
history. This has largely been accomplished through food.

This gastro-political project—termed a gastronomic revolution by many
(Fan 2013; García 2013; Matta 2016; Tegel 2012)—is anchored by the idea of
fusion, the mixture of cultural, racial, and culinary influences that combine to
produce high-end Peruvian cuisine that has allowed Peru to market itself glob-
ally with tremendous success (L. Fraser 2006). For instance, in the 2016 list of
the World's 50 Best Restaurants, three were in Lima;[3] similarly, in 2018, Peru
was named the "World's Leading Culinary Destination" by World Travel for
the seventh consecutive year. This culinary success has bolstered a collective
understanding of Peru as a nation "of all the races" (*de todas las sangres*), as
José María Arguedas, the famous Peruvian novelist and anthropologist, put it
in his 1964 novel. This is a hopeful reading of *mestizaje* (or racial mixture),
strikingly different from previous iterations that saw mixture (racial and cultural) as problematic and ugly. Culinary fusion has reframed Peruvian mixture
as beautiful, as possibility, as necessary for modernity. This has also implied a
reframing of indigeneity (at least in theory). Whereas just over a decade ago
Indigenous peoples were referred to as obstacles to modernity and faced brutal
violence, today, the story goes, they are valued as key protagonists in the gastronomic revolution that has allowed a national rebranding. This is why accounts
of Peruvian resilience and resurgence today also include discourses of inclusion

and tolerance. In fact, an essential part of the push toward culinary greatness is the explicit link made between the *aesthetics* of urban chefs (mostly male and mostly white) and the *ethics* of collaboration with rural (and primarily Indigenous) farmers and producers. Hand in hand with discussions of sustainability and biodiversity, dominant narratives deploy Peruvian food as a social weapon (Fan 2013; Santos 2012), one that has made possible the move from fractured to unified nation and insistently linked to the "beautiful fusion" of race, culture, and history. "Peruvian cuisine finds its source in love," Peruvian chef Gastón Acurio tells us. "The same love that made Lima [Peru's capital] the city of all the races" (2016, 98).

Importantly, a guinea pig boom has accompanied the Peruvian culinary revolution. Over the past decade, as demand for cuy meat soared, scores of small production businesses emerged, most of them owned or managed by working-class men trying to secure livelihoods. Every month there are several cuy production workshops (both on site and online) offered by private companies and state entities;[4] there are cuy-producer conferences at regional, national, and international scales; and the export of guinea pig meat has reached unprecedented levels. The head of one cuy-production business worked tirelessly to establish a national holiday for the cuy, and in 2013, the Ministry of Agriculture declared the National Day of the Guinea Pig, to be celebrated yearly every second Friday in October. The goal is to promote and increase guinea pig consumption in the country and abroad.

These two narratives—one about high-end fusion cuisine, the other about cuy production and small businesses—are two parts of the same story about national possibility, capital accumulation, and entrepreneurial success. Peruvian culinary fusion has cosmopolitan inspirations and aspirations, but it relies on traditional ingredients that authenticate the cuisine as Peruvian: quinoa, ají, alpaca, and, of course, the guinea pig. It is also a high-end culinary experience that relies on low wages in the rural countryside and peri-urban environments that, in Val Plumwood's (2008) terms, would be the "shadow places" that fuel the "nice places" of high-end restaurants and culinary festivals. In this chapter, I want to think about the particular significance of the cuy to the assemblages of race and nation in Peru. More specifically, I want to think about the figure of the cuy. Is the cuy a transformative force in Peru ("the cuy will save Peru") or just raw material for the gastronomic revolution? How does the guinea pig work—or how is it worked upon—to power the gastro-political machine, to support the move to reclaim sovereignty?

Thinking with Guinea Pigs: Histories and Representations

Figuration is about resetting the stage for possible pasts and futures.
— Donna Haraway, "Ecce Homo, Ain't (Ar'n't) I a Woman, and
Inappropriate/d Others: The Human in a Post-Humanist Landscape"

There is a fascinating figure in Peruvian popular culture called El Cuy. Drawn by Juan Acevedo, El Cuy was the protagonist of an eponymous comic strip that first appeared in 1979 in a famous left-wing paper, *La calle*. This was a time of military dictatorship in the country, when independent publications were prohibited. In other words, El Cuy was born a subversive, Andean (rural to urban) migrant figure. With the turn toward democracy, the emergence of Sendero Luminoso, and the eventual return of authoritarian rule, El Cuy moved through various iterations and publications, but as Acevedo writes, "always on the side of social justice and freedom."[5] Most recently, El Cuy has his (the figure is unquestionably male)[6] own blog and appears frequently in *El Comercio*, the Peruvian paper of record.

In his discussion of the birth of El Cuy, Acevedo tells his readers that he wanted to "create a character that would represent Peruvianness [*lo peruano*]." While this is a particular reading of Peruvianness—one squarely on the side of marginalized subjects and representing a particular masculine and "popular" subjectivity—the equation of the cuy with Peru is significant. The guinea pig has been an important figure in the Andean world for centuries (Archetti 1997; Morales 1995; Yamamoto 2016). The animal was domesticated by Native peoples at least three thousand years ago in the Andean region of South America, and cuy remains can be found in archaeological sites and in colonial art across the region. A famous example is the painting of the Last Supper hanging in Cusco's cathedral, where cuy is the main dish. Recent efforts to intensify cuy production claim the animal as quintessentially Andean and re-center the Andes in the Peruvian national (and subaltern) imaginary. Of course, there are long histories of Indigenous interspecies relations with the cuy that make possible this reclaiming of the cuy in the first place. And yet these histories, these ongoing relations, are all but made invisible in the current refiguring of the cuy. Tapping into anthropological discussions of *Perú profundo*, or "deep Peru" (Mayer 1991), the Andean origins of the cuy authenticate guinea pig production efforts as a *Peruvian* endeavor, as part of the broader national culinary project that, similarly, uses indigeneity as its anchor. But this is a move that both acknowledges and invisibilizes Indigenous labor with the cuy, a move that centers the past and extracts archeological evidence to commodify and appropriate. Consider the following passage from a well-respected manual for cuy production:

> The tremendous evidence found about the origins of the cuy in the Andes allows us to affirm, quite properly, that the cuy is *our* animal. In other words, he is from the Andean regions of Peru, Ecuador, Bolivia and Colombia, and—*as he is ours*—he obliges us and commits us to not only improve his exploitation and quality but also to create our own technology that deserves, in the future, our very best efforts. (Rodríguez et al. 2009, 27; emphasis added)

Similarly, a website advocating cuy production and genetic improvement states,

> The cuy was intensely consumed by the people of ancient Peru until the Spanish conquest that tried to do away with our culinary tradition, and incorporated instead meat from cows, pigs, and sheep in our diet. Since then, in a process taking place over centuries, people began to see the cuy with different eyes. There are people today who are horrified to see this rodent, or a piece of the animal, on a plate. Without a doubt, these people know nothing about history, and even less about good food. (Rico Numbela 2010)

This perspective relies on a blend of two familiar tropes that help reproduce species and racial hierarchies: a human-centered notion of "dominion" over the nonhuman world along with a Peruvian, nationalist reading of the country's Andean origins. Human dominion over guinea pigs is not only sanctioned by the familiar kind of theological arguments found in Genesis but also a sort of archaeo-national logic that gives Andean nations the opportunity—no, the obligation—to improve and exploit "our animal." Once again we see the familiar idea of "time and the other" (Fabian 2002): even as the cuy is rescued from the past and mobilized for the future, Indigenous peoples remain simply out of time; they are nowhere in the connection between nation and national rodent.

It is here that we might usefully return to Donna Haraway's (1992) idea that figuration sets the stage "for possible pasts and futures." The figure of the cuy makes possible a rearticulation of the past, of past laboring relations between "the people of ancient Peru" and this animal, in ways that skip over the present and instead offer new pathways toward modernity. But figuration, as I am using it here, is not simply a foregrounding of the figure and backgrounding of the flesh. Rather, drawing on Claudia Castañeda's work, I use figuration as a way to think about the articulations between "the semiotic and material" and, as Haraway argues, between possible futures and pasts. These kinds of relations (semiotic, material, and temporal) work to produce specific "figures" that necessarily bring forth connections to particular worlds. As Castañeda writes,

> Figuration entails simultaneously semiotic and material practices. This concept of figuration makes it possible to describe in detail the process by which a concept or entity is given particular form—how it is figured—in ways that speak to the making of worlds. To use figuration as a descriptive tool is to unpack the domains of practice and significance that are built into each figure. A figure, from this point of view, is the simultaneously material and semiotic effect of specific practices. (2002, 3)

Castañeda's focus is on the figure of the child, a being who, she argues, is always in the making, always potentiality rather than actuality, a temporally ambiguous being that is full of futurity but clearly anchored in a genealogical past (2002, 1). We can think of the cuy in similar ways. The cuy's power lies in its—and I flinch at my complicity in figuring the cuy as an "it," but this too is part of what we must explore—yet-to-be-ness but also in its always already Peruvianness. This is precisely the kind of figure that is needed in a post-crisis moment of nation building, something that can anchor the imagined community of Peru in a *longue durée* of being and belonging to a place and environment yet can also lend itself to the cosmopolitan and modern innovations of tomorrow. And yet, it cannot be mere symbol but must also be edible flesh. As export commodity, ingredient, Andean body, and national symbol, the cuy is quite literally on the move, even (especially) beyond its own death. In both a sacrificial and symbolic way, the cuy powers many practices and projects.

It is worth pausing here to consider how the cuy is central to thinking about "how nature works" (and is worked upon) in the Peruvian context. The figure of the guinea pig does great work—as symbol of future possibility, as Indigenous nature transformed into "modern," high-end cuisine—in reframing Peru, a society once on the brink of internal destruction, as a sovereign nation. This reframing is intimately linked to indigeneity and is inextricably tied to long histories of elite formulations of the "problem of the Indian" and the need for

Indian pasts to authenticate Peruvian futures (de la Cadena 2000; García 2005). Kate Soper tells us that "'nature' is the concept through which humanity thinks its difference and specificity" (1995, 155). Moving in a similar, but perhaps more radical dimension, Bruno Latour suggests that we are better off without "nature" since this "is what makes it possible to recapitulate the hierarchies of beings in a single ordered series" (2004, 25). Although I agree with Latour's normative push against hierarchy and human exceptionalism, I suggest that before we abandon "nature," we should continue to explore how it works discursively and materially to sort and rank, even as we disagree with the ethics of that sorting and ranking. As I will argue below, gastro-politics in Peru redraws naturalized racial hierarchies in novel and familiar ways; it can be understood only in the context of Peru's colonial legacies, where nature and indigeneity are coproduced. Native landscapes and peoples come together in the figure of the cuy, evoking worlds of the so-called dirty, backward Indians who must be cleansed; that is, distanced from their "natural" environments so that they can be brought into a modern and cosmopolitan nation anchored by romanticized and sanitized forms of indigeneity. Asking "how nature works," as the editors of this volume encourage us to do, I am struck that "nature"—pace Latour—works very much in the same mode as race. Stuart Hall puts it clearly: "One must start . . . from the concrete historical 'work' which racism accomplishes under specific historical conditions—as a set of economic, political and ideological practices, of a distinctive kind, concretely articulated with other practices in a social formation" (1980, 338). To put it simply, "Notions of race and nature . . . work as instruments of power" (D. Moore, Kosek, and Pandian 2003, 8). These forms of power work as a complex of material and semiotic practices; in other words, as figuration.

Moreover, the figure of the guinea pig is central to elite formulations of the labor involved in transforming Peru into a global commodity, into a sophisticated culinary destination. But this is work that centers nature only as it is sanitized and made palatable through the "labor of love" performed by chefs as they remake a rodent into airy cuy puffs. Indeed, nature is only valued as worthy of admiration when it is worked upon by celebrity chefs, such as Virgilio Martínez (2016), who "cooks ecosystems" by "discovering previously unknown ingredients" and transforming them into edible works of art, each embodying a piece of Peruvian biodiversity (see also Jeter 2017). In the context of Peruvian gastro-politics, then, nature only works through the lens and labor of celebrity chefs. Put another way, gastro-political elites in Peru put nature to work in ways

that cleanse and erase the traces of Indigenous labor that made possible particular understandings of "nature" to begin with. For instance, through their
culinary labor, elite chefs make the cuy "work" in ways that once again render
Indigenous peoples and guinea pigs into timeless nature, awaiting "discovery"
and then transformation into beautiful food—tradition and modernity on the
plate. This move also erases centuries of interspecies relations and labor. As
one of the editors of this volume noted, we could read this as a kind of labor
struggle over whose labor can accrue value. Indigenous activists in Peru make
a similar move in pointing to this as an ontological struggle, one that highlights
the violence involved in extracting cuy from specific relations and refiguring
the animal as fusion cuisine.

In the following sections, I explore some ways in which the cuy is figured,
and the various worlds the animal, the *figure*, makes possible and inhabits.
Fusion, as we will see, is a central element in the transformative act of figuration. In an almost Lockean way, gastro-political elites make nature work and
also make it their/our own by mixing their labor with both material and semiotic elements (Locke 1980). Just as Locke has been criticized for the ways in
which Indian land, labored on by black bodies, somehow magically produces
white property (Wolfe 2016), we might similarly observe a colonial alchemy
in which Indigenous dispossession and even disappearance creates the very
conditions for capital accumulation. Figuration resignifies the labor of other
human (Native peoples) and nonhuman (guinea pigs) beings.

Interlude: Indigeneity and Cuy at Mistura

The old woman made her way through the throngs of people, looking confused
at times but persistently calling out "*¡Cuy! ¡Rico cuy!*" No one looked at her. All
around her, people moved happily from the Gran Mercado (the grand market)
to food stands to beer gardens. They looked at their maps to decide where they
might go next. They stood in line for the public bathrooms. They listened to
musical performances or lectures about composting and the value of organic
produce. Children ran past her, kicking up dirt and dust, almost knocking her
down. I watched as she put down her basket to rearrange her pollera (traditional Andean skirt) and her hat. She picked up her basket, looked around again
as if looking for someone, and then a bit more loudly this time, called out once
again, "*¡Cuy! ¡Rico cuy!*" She took a skewered fried guinea pig out of the basket,
waving it around a little before putting it back in.

Suddenly, as if out of nowhere, three security guards appeared. They were dressed in black and brown; pink-and-yellow tags indicating they were Mistura guards hung from their necks. They surrounded her, grabbed her from the arms, and almost lifted her off the ground. The woman looked more angry than scared. From where I was, I could not hear what they said to her or what she replied. But I could see they wanted her out of the festival area, and they were taking her in the direction of one of many exits. Just as suddenly as the guards had appeared, two younger women arrived, waving their arms and yelling out. I had moved closer by then, and I could make out a few words: "Huanta," "*productora*," "*con nosotros*" (Huanta, producer, with us).[7] After a brief exchange, the guards let the woman go, and the two women surrounded her, held her, and led her away from the guards and toward the Gran Mercado, where they disappeared into the maze of products and food stands.

Figuring Race and Nation

The vignette above describes a scene from Mistura (literally, "mixture"), widely recognized as the most important culinary festival in the Americas and the most significant yearly event in Peru (Perez 2011). In 2008, the website for Mistura noted that

> Mistura is a cultural party, where the best comes together from our gastronomy, haute cuisine, cooks, pastry makers, neighborhood bakers, regional cooking. Also, peasant farmers, fishermen, and cattle breeders visit us in their traditional costumes and with the most typical produce from our Amazon, Andes, and Pacific Coast. They all find in Mistura the perfect occasion to joyfully celebrate our gastronomy and grant recognition to the hard work that each person plays in the gastronomic chain. . . . We all gather at Mistura to celebrate our tradition, creativity, identity, and diversity. . . . Mistura #*Somos Todos!* [We are all Mistura].[8]

Mariano Valderrama, general manager (and former president) of the Gastronomic Society of Peru (Apega), adds that Mistura is "one of the brands that best represents *la peruanidad*" (2016, 14). This festival has been one of the central components of a broader national branding campaign: Marca Perú, or "brand name Peru" (Cánepa 2013). One initiative of Marca Perú was the 2015 online #MásPeruanoQue ("More Peruvian Than . . .") campaign. This campaign, a collaboration between Apega and PromPerú, was an effort to expand the marketing of Peru in the world: "Peru is a great brand. A country that makes you fall in love. . . . Being Peruvian does not have geographic limits. Loving Peru does not require passports."[9] The website goes on to note the importance of the global love and promotion of Peru ("If the world is proud of being Peruvian, then how can we Peruvians not be proud of ourselves?") and showcases videos of French, Puerto Rican, and Panamanian individuals who promote Peruvian food and music. As part of this campaign, posters went up in certain neighborhoods in Lima (high end, tourist friendly) depicting different icons of Peruvianness. Posters with an image of the cuy (*Más Peruano que El Cuy,* "more Peruvian than the cuy") were, not surprisingly, some of the first to go up (figure 7.1).

As I have argued elsewhere (García 2013), Mistura is a particular kind of national performance, a key part of a broader national project that presents a beautifully mixed and tolerant nation to the world, a nation anchored by its Indigenous past and its histories of migration, all brilliantly exploited in this new moment of nation branding and culinary greatness. Chef Acurio, perhaps the most important spokesperson in this moment, puts it this way:

> We are convinced that our cuisine is the fruit of a long, tolerant relationship among people and a treasure trove of ingredients that is the result of centuries of dialogue between our ancestors and nature. . . . It is also the cuisine of a country to which different peoples . . . migrated over the last 500 years. All brought with them their nostalgia, customs, and products, which were beautifully assimilated into an example of unique tolerance. The result is a Peruvian cuisine that infuses a little of each of those peoples into each bite, transforming it into something new, something Peruvian. (2015, 8–9)

And yet, woven throughout this celebration of diversity is a clear message that the gastro-political project in Peru—including Mistura—is in fact about managing difference, about repositioning and authenticating difference in order to make it safe. The attempted removal of the old woman with her cuy offers a glimpse of the ways difference is managed and the significant role that Native producers play in this project. They are touted as "the protagonists of Peruvian cuisine" (Moseley-Williams 2013) and are expected to perform a particular kind of authenticity at markets and festivals by wearing traditional clothing—always colorful, always bright—and presenting "beautiful, happy faces" to

Figure 7.1. Poster from
#MásPeruanoQue campaign at Mistura.
Photograph by author, September 9,
2015, Lima, Peru.

the public.[10] Native Peruvians occupy a seemingly paradoxical position in this
national story: they are presented by gastro-political elites as both protagonists
and crucial background material; they are necessary to anchor and authenticate this new, modern, Peru, but they must be in their place, both physically (in
Andean or Amazonian spaces) and historically (in the past). The woman in the
vignette above, selling cuy prepared in the traditional manner, disrupted this
staged national performance. She—and the cuy—were entirely out of place.

Like Mary Douglas's famous definition of dirt as "matter out of place"
(1966), the woman and the cuy—skewered, with visible teeth, nails, and hair
still present—were seen as dirty figures who did not belong and should be
removed from the festival, from this so-called national celebration of diversity. The cuy figured a particular past, one where cuyes were raised in homes
on dirt kitchen floors, not in large confinement facilities; it remembered the
ugly, dirty, deep Peru that gastro-political elites have worked so hard to remake,
to erase. As Rudi Colloredo-Mansfeld (1998) has argued in his work on the
northern Andes, the creation of bourgeois subjectivities and modernities in
the Andes has since the nineteenth century "hinged on problems of hygiene"
(1998, 187). Building on the insights of Michel Foucault (1990) and Ann Laura

Stoler (1995), Colloredo-Mansfeld reminds us that "colonizers elevated hygiene into a gendered and racial 'micro site' of political control. It provided a context for appraising racial membership and designating 'character,' 'good breeding,' and proper rearing" (1998, 187, citing Stoler; see also McClintock 1995). It is not surprising that hygiene and concern over "dirty Indians" continue to preoccupy the designers of the new "beautiful fusion."

Mistura's message of inclusion notwithstanding, when the festival proclaims that "we are all Peru," it means both a particular "we" and also a particular "Peru"; it includes those who accept a particular projection of the nation, who abide by certain expectations of hygiene and compliance, and who know their place in the hierarchies of emerging networks of local-global capital flows. When I spoke to one of the festival organizers, Luis Ginocchio, he was very aware of the Andean woman trying to sell her cuyes. "She was selling her cuyes like a street vendor [*como ambulante*]!" he exclaimed, and continued, "You can't do that; you can't sell out of place; you have to be in your place; the right place." Ginocchio was referring to the vendor stalls behind which each producer was expected to stand. He told me the incident demonstrated the work that still needs to be done to "train them," and that this was an example of how easily all the work they have done to promote cuy and other Native products can be undone. He said, "We don't eat [the cuy] whole; certainly not with its head still on. People don't want to look at the face, especially tourists or Peruvians from a certain class. Her actions are taking us backward." Mistura—and the gastronomic revolution it represents—can be read as a crucial component of what Norbert Elias (1994) famously described as the civilizing process. The Peruvian brand, the marketing of nation, relies on the beautification of that nation, and in elite terms, this has everything to do with taste, with a particular presentation of national culture (and cuisine), of that which authenticates. Raúl Matta notes that making Peruvian ingredients palatable necessitates the ability to "neutralize their unworthiness, their 'Indianness,' and their lower-class characteristics" (2013, 6). This attempt to "neutralize," repackage, and make Indianness consumable is especially evident in the festival organizer's comments above.

It is worth remembering that the gastro-political project markets itself as in part responding to the violence and disorder of the recent past. During the internal conflict, the intensification of internal migration, the emergence of informal settlements and shantytowns surrounding Lima, and the increasing presence of Indigenous and peasant bodies in coastal Lima (what some call the Andeanization of Lima) greatly exacerbated racial tensions in the country (Matos Mar 2004; Martuccelli 2015; Nugent 2012). Moreover, the state's authoritarian approach cloaked the country in fear, and political and economic instability plagued Peru. For Peruvians, things were out of place. Indigenous peoples, many literally displaced from their lands and homes, often talk about this moment as one where the world was turned upside down. In Lima, the sense of a lack of order permeated everything. The arrival of Sendero Luminoso to Lima was announced by the bodies of dead dogs hanging from lampposts. With the war no longer only in the Indigenous countryside, urban residents of Lima found themselves craving order at any cost. This is perhaps why 80 percent of Peruvians supported Alberto Fujimori when he declared a self-coup in 1992, closing congress and the courts (Tanaka 1998). In this context, to return to the animal, guinea pigs—quite clearly linked to Indigenous spaces, and increasingly present in Lima as part of migrant households—were very much seen as out of place. They were disparaged as "Indigenous rodents," food that belonged in Andean households, not in Lima.

It is striking then, that by the early 2000s, Peru had turned from a place of terror to a tourist haven. Suddenly, the country was now famous for its 2010 Nobel Prize in literature, an Oscar nomination for best foreign language film in 2009, and increasing economic stability and development. This is also when food enters the scene. Capitalizing on the idea of fusion, and keenly aware of the backdrop of violence, racial antagonisms, and political turmoil, Acurio and others sold Peruvian fusion cuisine to Peru and the world as the country's savior. This was a culinary movement made possible by the new moment of peace and prosperity. The Truth and Reconciliation Commission noted that Sendero was made possible because of the neglect and racism that was part of everyday life for most Indigenous Peruvians. People had nothing to lose, so they supported the revolutionary organization. Thus, these populations needed to be "brought in," or be made to feel a part of the nation. Working with Native producers, making space for them (and the products associated with them) at the table, is a crucial part of the story. So, of course, is the cuy.

To go back to Castañeda's notion of figuration: how is the cuy given particular form—how is it figured—in ways that speak to the making of worlds? Both the Andean woman and her cuyes were out of place in Mistura. The lesson was clear: difference and mixture must be managed. Indigenous and cuy bodies are acceptable in spaces that were previously off limits, but only as long as those bodies come in particular forms and follow particular rules. This may certainly be an improvement over the times of terror and fear, but the racial orders that generated much of the violence of the 1980s and 1990s have found themselves remarkably adaptable to the new slogans and spaces of mixture and tolerance. Lima may be far from the days when dead dogs hung from lampposts, but the arrival of capital-friendly guinea pigs may yield other disturbing lessons.

"From Savage to Sophisticated": Figuring Haute Cuisine

Rodent never tasted this good.
—Simeon Tegel, *Guardian*

The epigraph above comes from an article in the *Guardian* about "Peru's fantastic food revolution" (Tegel 2012). The article begins by noting that for most tourists, local Peruvian fare "used to mean roast guinea pig, nibbled en route to Machu Picchu." Tegel continues, "The idea of coming here specifically to eat would until recently have elicited bafflement, if not derision." And after describing the recent rise of Peru (and more specifically, Lima) as the gastronomic capital of the Americas, the author ends his piece by telling his readers that if they are "set on eating guinea pig," they should "try the version at Astrid y Gastón, which the menu describes as having visited Beijing before travelling to Shanghai. Rodent never tasted this good." Astrid y Gastón is a renowned restaurant, the first in Peru to be recognized as among the World's 50 Best Restaurants. Housed in the renovated Casa Moreyra, the three-hundred-year-old main house of the old hacienda of San Isidro (one of the wealthiest districts in Lima), it offers both tasting and à la carte menus. The famous guinea pig dish the *Guardian* referenced is Peking Cuy (*el cuy pekinés*), listed by an author in the United Arab Emirates *National* newspaper as one of the "ten dishes to try before you die": "Traditionally the 'cuy' was deep fried. . . . But, fortunately for pet lovers, Acurio . . . has disguised it here as Peking duck, serving it sliced up into chunks with a rocoto pepper hoisin sauce. Once it's wrapped up in a purple corn pancake, you won't even think about Fluffy" (Bartley 2013).[11]

Recalling Castañeda's insights about the material and semiotic dimension of the figure, we see that the power of the figure lies in both what it makes possible—and erases. As she notes, the "figure is a resource for wider cultural projects" (2002, 2). Here, the figure of the cuy evokes the authenticity of indigeneity, but it is important to note just how much labor goes into transforming and obscuring the cuy in ways that accord with lingering colonial notions of hygiene and the cosmopolitan aspirations of the gastronomic elite. The animal is appropriated by elite chefs who transform this otherwise dirty, Indian rodent into the key protagonist in tasting menus at high-end fusion restaurants. The cuy is once again essential for the transformation "from savage to sophisticated,"[12] but the animal must remain hidden.

I should be clear that the gastronomic boom involves many more animals, ingredients, and strategies than the ones described here regarding the cuy. Nevertheless, the cuy—as animal—is in many ways the indispensable figure of the nouveau Andean turn as it anchors this gastro-political project in a particular understanding of Peru as an assemblage of nature, history, and race. As a symbol of a millenarian history of domesticating the natural world of the Andes, the cuy, along with alpacas and llamas, points to a long history of Andean knowledge and civilization. Nevertheless, the discomfort motivating the transformation of this traditional food animal betrays the familiar slogan of *indigenista* projects, "Incas sí, Indios no" (Méndez G. 1996). In other words, a gesture to the glorious Indigenous past is important but so is a thorough scrubbing of the smelly, dirty, and ugly Indigenous present in all its human and nonhuman forms. This is once again a project of the management of difference. For the cuy to make the voyage from its "natural environment" (in the Indigenous rural countryside or perhaps the markets of migrant communities like Paterson, New Jersey), it is only appropriately placed in high-end restaurants when it is transformed—turned into cuy ravioli or exoticized into the cosmopolitan Peking cuy—into an animal that could only exist in the contradictory temporalities and geographies of the gastronomic boom.

Conclusions

It is important to note that elite appropriation of Native knowledge and labor is only part of the story. There is no question that some Indigenous producers benefit from this rediscovery of the cuy, not a small thing given rampant inequality and poverty in the country that impacts primarily Indigenous peoples and internal migrants. Moreover, the recognition of the cuy's significance to national development seems to have elevated the animal to a different place in the national imaginary. It seems to have somewhat lessened the disdain elite and middle-class Peruvians had toward guinea pigs, probably in part due to the association of the animal with rural and Indigenous worlds. Beyond elite tasting menus, then, there is a space that has opened up for middle and popular sectors to harness their aspirations to this Andean animal, as the dozens of cuy-production businesses in the country attest to. It is also worth noting the gendered nature of some of that labor, as it is usually men heading those businesses and the female cuy *reproductoras* (breeders) who—through multiple impregnations and separations from her young—stand at the center of the guinea pig boom.[13]

To go back to the focus of this volume, in this chapter, I have suggested that much labor goes into making (and unmaking) the cuy. As edible flesh, it provides the seemingly "raw materials" for the mixing of labor and nature required to make fusion an economically productive activity. Just as in Lockean understandings of nature and labor, the animal is made for labor and is figured in ways that make invisible the long histories of Indigenous interspecies labor with cuyes. Additionally, the figuration of the cuy does the spatio-temporal work of connecting multiple pasts and presents. Similar to the way tea in India serves as a temporal bridge between plantations of the "past" and contemporary monocultures (Besky, this volume), the cuy serves as a bridge between Indigenous pasts and cosmopolitan futures, between rural and urban Peru, and between a society characterized by internal war and terror and the hope of a society of beautiful fusion and harmony. This is not to glorify the biopolitical and culinary work that has recruited this sentient being to these large human-centered causes. In many ways, the cuy is the cannon fodder of the gastronomic revolution. Moreover, the optimistic claims of inclusion and harmony that come with the "beautiful mixture" of the culinary boom serve to distract observers from its unquestionably exclusionary and hierarchical dynamics. The cuy, as an inhabitant of Indigenous worlds and an ingredient in cosmopolitan haute cuisine, is crucial to the postwar remaking of Peru, as once again, a modern nation roots in ancient soils.

Notes

1. I interviewed Chauca in her office at the INIA in Lima in 2009, 2011, and 2012. The discussion here draws from these interviews. Zaldívar, Chauca's colleague and husband, was killed by Sendero Luminoso (the Shining Path) while he was at one of their field sites in Ayacucho, the highland region where this Maoist insurgent group first emerged and that was hardest hit during the political violence of the 1980s and 1990s. For more on the history and impact of this period of violence, see Degregori (2012) and Stern (1998).

2. All translations are by the author.

3. See https://www.theworlds50best.com/list/past-lists/2016.

4. These workshops, offered to those interested in developing their own cuy production business, target working-class men and women.

5. See text on "El Cuy" tab (https://elcuy.wordpress.com/about/). Accessed January 15, 2019.

6. El Cuy is gendered male, as is the case more generally with various contemporary figurations of the cuy.

7. Huanta is the name of a town in the highland region of Ayacucho.

8. This quote came from the section *nuestra historia* (our history) on the website for Mistura (mistura.pe) in 2008. This quote is no longer available on the website; notes from the 2008 version of the site are on file with the author.

9. The quotes from this section come directly from the website for this campaign: http://masperuanoque.pe/#campaña. Accessed January 19, 2019.

10. This discussion and quotes following draw from an interview with Luis Ginocchio (member of the executive council of the Gastronomic Society of Peru) in Lima during the Mistura 2015 festival.

11. Once again, we see clearly the way the labor of elite chefs is what makes cuy work in this context, with no consideration of the history of Indigenous-cuy relations or the labor of Indigenous peoples in domesticating this animal.

12. This was a phrase I heard several times during early field research for this project, mostly from elite Limeños participating in this culinary movement.

13. Moreover, as we consider the figure of the cuy in the economic work of the culinary boom, it is striking how the cuy is more often than not gendered as male. This is perhaps not surprising as many visions of modernity, of both right and left, share the same masculinist view of the subject of history that moved teleologically away from not only "nature" but also from ethnic and racial particularity (Saldaña-Portillo 2003).

CHAPTER EIGHT

Industrial Materials

Labor, Landscapes, and the Industrial Honeybee

JAKE KOSEK

Beekeeping is the ugly stepsister of industrial agriculture; it does not play the leading role but is essential for the story. In fact, much of modern agriculture would not exist if it were not for the modern bee's mobility, its industriousness, and its hive, all of which allow for the pollination of vast fields of production. But this industrious and industrial bee is not the same beast it was a few hundred years ago, and the largest and most consequential factor in its physical, social, and spatial remaking has been its role as worker in industrial agriculture.

To be a beekeeper and to make keeping bees economically viable (though barely) is to submit to the production regimes of industrial agriculture; its sheer size, scale, and industrial configurations are the uncompromising conditions of modern beekeeping. But at this abstract level, the form of industrial beekeeping is still vague and amorphous. What I hope to demonstrate in the details of this chapter of the volume are some of the material configurations and consequences of the political histories of industrial agriculture. While bee deaths still make headlines once a year (statistics show that their population has dropped 30–50 percent annually since 2006), the statistics tells us little of the conditions of those deaths or, perhaps as importantly, the conditions of those that survive.

While I started out understanding the current conditions of bees as the result of structural economic forces, the details of the hive, the personalities and styles of beekeepers, the unpredictability of bees and mites, and chemical interactions complicated this formulation. While these economic conditions are central to this story, whenever I reduced them to the effects of human agency, the obdurate materials became vibrant; living beings became animated in unpredictable ways that frustrated my structural accounts of apiary histories. I realized that I was reducing history and politics to human agency so that objects became simply the products of the political economics of their making.

While these seemingly disparate political histories of keepers and bees conform to economic standards and logics of work and labor, they also have helped to make and change them. Physical landscapes, standard pallet sizes, weight loads, and federal regulations have all shaped the bee and are all moved by its possibilities and limitations.

The bee is moved too; it grows to fit, in part, the complex relations that materialize in its body: its size and shape, its seasonal rhythms, the very quality and chemistry of the nectar it desires—all are transformed through the conditions of industrial agriculture. Together with material objects, intimate practices of management, and the bee's desires, the bee's work becomes part of industrial agriculture's modern biological infrastructure. There are many dimensions and consequences to the bee's role in industrial agriculture: its mobility, its transformed diets, its vulnerability to exposure to a cornucopia of chemical concoctions and beyond. So while bee deaths are not predictable and do not have a single cause, the fact that they are often deemed inexplicable is frankly a blind abnegation of the conditions of the modern bee and the politics of its making.

This is not new or news for industrial-scale beekeepers who haul thousands of hives even more thousands of miles in an effort to pollinate a system of agriculture where the bee can no longer survive in the boom-and-bust flowering and pesticide-laden landscapes that produce almost all of our food. The rhythms of the seasons and the seduction of nectar are blended with the production of an industrial landscape, the socially necessary labor time of both the bee and the keeper. Jeff Anderson, owner of a fourth-generation beekeeping operation with a thousand hives, does not need to be reminded that the bee's health is in some respect overdetermined by the histories of its making, an industry that has turned the honeybee and its keepers into a modern form of production in order to survive. My interview with Jeff, like many interviews meant to take just an hour, stretched into an all-day event, beginning in his kitchen, then moving into his pickup to visit hives, then out to meet other keepers at a local diner, and then back to the house for dinner and to look at photo albums. What I learned from the long conversations on this day is that if we look deeper into the hives and their desires and the changing practices that Jeff lives, with his seasonal and endless migration, we can see that the fragments of the beekeeping industry's complex infrastructure tell us much about not just the history of the reorganization of the industry but also the making of the modern working bee. An honest epidemiology of the bee must start not just with the bee but with its modern materializations alongside the keeper and the conditions that it now occupies.

A good place to begin to explore the contemporary conditions of the honey-bee is in the Central Valley in California, where twelve thousand square miles of almonds bloom almost all at once in February. It is a stunning sight. The individual light pink blossoms on each branch of every tree are impossibly delicate, yet they are situated in a collective bloom of such enormity and seemingly perfect rows; the picture produces a confusingly intermingled awe of the beauty, enormity, and order of the industrial agrarian landscape. Almonds are only one crop of many in this landscape that bees pollinate, but they are the most lucrative. For this crop alone, over 1.5 million hives—60 billon bees and over 60 percent of the total living honeybees in the country—are trucked into California from around the country. At the current growth rate, there will need to be 2.3 million more hives to pollinate the almond crops alone by 2020; that is the current total number of hives in the country.

California almonds require the largest migration of pollinators in the world, which are brought in from forty states from late January to early March to facilitate arboreal promiscuity (Bishop 2005). The price of pollination used to be about $10 per hive in the mid-1970s; now, because of disease, the hive shortage, and the further concentration and size of fruit and vegetable production, the price is around $170–$185 per hive in 2017. There are full-time professional bee brokers who make deals between almond farmers and beekeepers to determine the size of the bee colony through the number of bees and brood per hive, the exact timing of when the hives are placed in the orchard to maximize pollination, and exactly when they need to be removed (so as not to get doused with direct spraying). The process of contracting with the orchardist is sometimes informal and often contentious. Sometimes it is based on handshakes and long-term friendships; at other times it is based on strict guidelines determining the number of bees and inspections of colony size and brood patterns (the brood density and quantity indicate the future colony's strength). Sometimes you just do not get paid.

One hundred and seventy dollars per hive may seem expensive, especially when we are talking about 1.5 million hives, but California almonds make more money than the entire California wine industry; the state produces 80 percent of the global almond supply, almost a billion pounds a year. Without commercial bees, the total would be less (some say a lot less) than fifty million pounds, about one-twentieth of current production. There are other vectors for pollination—some native bees, the occasional bird, wind, et cetera—but by bringing bees in, a grower can raise 30 to 40 pounds per acre to upward of 2,500 pounds

(Nordhaus 2011). One look at the massive quantity of blooms and it is clear that the occasional pollinator would never be adequate to pollinate each one. As a result, between 15 and 20 percent of the per-acre cost goes to pollination, about the same as irrigation, on average (Harvey 2012). Bees are central to the $3 billion-a-year industry; as Neal Williams from the Department of Entomology at UC Davis states, "Without the honeybees . . . the [almond] industry doesn't exist" (Lifsher 2012). Joe Macilvaine, president of Paramount Farms, a forty-seven-thousand-acre almond farm, says, "We need them to be reliable, and we need them at the right time. . . . Lots of things can reduce almond yield—weather conditions, drought, insect infestations. But if you don't have the bees, you never get to begin" (Lifsher 2012). With the declining population of bees and the growing acres of almonds, the price of pollination has increased to three times the cost it was less than ten years ago, to the point where in 2012, California growers spent about $250 million on bees (Lifsher 2012).

At present, honeybees pollinate over one hundred other crops in the United States. Almost all of our favorite fruits—apples, blueberries, cherries, pears, avocados, peaches, and melons—as well as vegetables like squash, cucumber, broccoli, onion, eggplant, and garlic, in addition to almost all citrus, are commercially dependent on the honeybee. The US Department of Agriculture (USDA) estimates that the honeybee is responsible for $20 billion in domestic income from direct pollination alone (USDA NRCS 2014). This number, and a belief in the importance of pollination for high crop yields, has steadily grown in the last century. This is partly because studies of crop yield have raised awareness, and partly as a result of farms over the last century needing to introduce outside pollinators. Regardless, the industrialization of farming as well as that of beekeeping, first for honey and later for pollination, is profound.

As I alluded to above, it is not just that bees pollinate the crops: the organization of modern industrial agriculture would not be possible if not for the honeybee. Carey McWilliams, one of the most famous chroniclers of the history and politics of industrial agriculture, has aptly said of the industry in California, "a new type of agriculture has been created" (2000, 5), and it is directly at odds with the agrarian myth of family farms producing both abundance for all and the qualities and character that form the backbone of American democracy. It is large scale, input and production intensive, and based on a single crop (commonly referred to as a "monocrop"). McWilliams and others point out that industrial agriculture in the United States requires land monopolization, specialization (monocrops), and large quantities of cheap and mobile labor. To

cultivate at this scale and to manage the labor and profit requires, as in all industries, a reorganization of the time and space of production. It is well recognized
that the political geography of the landscape that melds the hydrological cycle
with the political histories of water law and hydro-engineering, the quality of
the organic content of the soil and the chemical charge of nitrogen and surplus capital needed to improve it, the length of the growing season and the
economics and violence of cheap labor, and contradictory racialized immigration laws have all helped make modern industrial farming in the Central Valley
possible.

Much overlooked, however, is the labor of the honeybee—its mobility and
manageability, its qualities as a pollinator, and its ability to live in large social
communities—that has made the concentration of monocrops biologically and
economically viable. While there are many thousands of species that pollinate
plants, the fact that bees can be moved to any location en masse is critical. As
lifelong beekeeper Steve Ellis says, "You cannot simply put native bees in the
millions on the backs of trucks and move them around the country. . . . Honeybees are the only viable solution given the way modern agriculture works today.
There really are no alternatives."[1] The almond industry has searched for alternatives, from bumblebees to self-pollinating almonds, but it is far from making
either of them a viably cost-effective and efficient alternative to bee-pollination
management. So, for now, industrial agriculture relies on the materials and
socialities of bees and keepers so it can be scaled at such a level that farms
stretch literally for miles and bloom all at once. It also relies, in particular, on
temporary labor (both human and apiary). When a farm only grows one crop,
there are times when it will need large numbers of employees for planting and
harvesting, and other times during the changing cycles when the employees
will be idle. It is much more profitable for the farmer if that labor pool is there
when needed and gone when not so the farmer has no obligation to the worker.
The structural effects of industrial agriculture in some ways impact both immigrant field workers and migrant honeybees. Pollination is only needed at specific moments, and then bees become a burden to keep alive. In fact, even if
bees stayed on site and could somehow survive the toxicity of pesticides and
herbicides, and the dearth of weeds in and around the fields, they would simply
starve because the bloom is concentrated at one specific moment.

Unlike many of the other agents of this volume, from mushrooms to orangutans, there is little debate about whether bees work. In fact, the bee is the
quintessential worker in modern Western history. The hive was studied and

cited by labor optimizers as a model for the factory, and bees were theorized as workers by economists from Bernard Mandeville to Karl Marx to Friedrich von Hayek and beyond. Female bees other than the queen are called "worker bees" and are, as I describe elsewhere, the animal most commonly invoked to naturalize labor relations. This history, along with the fact that bees are subject to the same structural conditions of industrial agrarian production, makes it that the structural conditions of migrant farmworkers are not accidentally similar to those of bees: poor living conditions, forced migration, exposure to chemicals, labor subcontracting, disposability, and making the living conditions of the labor force a social problem while making the surplus value private. Similarly, these necessities are not "choices" of good farmers or bad ones but the structural conditions of industrial agriculture, as McWilliams (2000) and many others have made so clear.

To say immigrant laborers and migrant bees are related is not to say they are the same; clearly, they are not. But to say that the bee is simply a metaphor for labor relations is to miss that the labor relationship has legal, social, and economic infrastructures that are codified in fieldwork, forced migration, chemical exposure, et cetera, that similarly conditions the labor exploitation of both bees and humans. Bees' work is not about consciousness in the narrow sense. Do bees work as an enabling force for productive action? They clearly do. Do they become working-class subjects with traditional ideas of consciousness through this action? No, they clearly do not—but stopping the analysis there misses so much of the consequential animacy and politics of agrarian production.

Industrial agriculture enables high yields with minimum labor because standardization and simplification of crop diversity allow for a large number of repetitive tasks and the possibility of greater mechanization. With chemical fertilizers and pesticides, mechanization thus becomes even more possible because variation in soil and pest populations can be managed through these inputs. The result is a very high-yield agriculture and high-profit industry that is heavily mechanized and chemically dependent, with an intense but seasonally variable labor force. As many have pointed out, this arguably makes economic sense, narrowly defined, but has a high ecological cost as diverse systems, geologies, and species are bent to the simplifying logics of a particular economics (McWilliams 2000; D. Mitchell 1996; Guthman 2004). Though its extent and effects vary widely, this combination of economic logics of efficiency and industrial agriculture has done more in the last century to transform global landscapes than anything else has. The ubiquity and scale of contemporary agricultural production makes the types of sustainable farming that deal with the broader ecological costs and social justice issues economically unsustainable.

The dangers of industrial agriculture are widely known; focusing on consumption politics, technological fixes, and education as ways to address industrialism may be important, but their effects are limited. There are structural constraints that compel those involved in agriculture to act in ways that may make economic sense, even at the cost of the long-term viability of the resource being used. Moreover, they bracket deeper systemic change and separate the politics of justice and the economics of production from the far more palatable feel-good solutions, without thoughtful and meaningful transformation of the food industry.

The practice of moving hives is not new, but industrial agriculture has captured and transformed it—and the honeybee. There are sporadic accounts of moving hives in the United States before the industrialization of agriculture, but it was not until the late 1800s and early 1900s that it developed on a larger scale. Most beekeepers did not produce more than a ton of honey per year. It was not until the rise of really large-scale monocrops between World War I and World War II, particularly during and after World War II, that agricultural production was transformed through the fusing of industrialization and the war machine. The intensity and scale of agriculture and beekeeping blossomed.

While early on the beekeeper helped produce higher yields for the farmer, who might have paid a nominal fee, the real benefit in moving the hives was a larger crop of honey for the beekeeper. Moreover, the demand for pollination was much less difficult to meet after World War II because there were more than twice the number of hives than today—between 5 and 6 million compared to today's 2.5 million, which is much closer to the 1940 census of 2.3 million (USDA 1940; USDA NRCS 2014). The increase of bees and a full doubling of the hives in the United States during World War II were largely the result of a government push to increase wax production, which was used to cover munitions to protect them from corrosion. There was also a push to produce more honey because sugar was tightly rationed. After the war, sugar rationing declined and prices crashed, putting many beekeepers out of work.

Because of declining beekeepers and new disease threats, the US government started a honey subsidy program with the Agricultural Act of 1949, which protected beekeepers from the vicissitudes of the markets. At the time, beekeepers produced honey and pollinated crops. The stated reason for the promulgation of the honey subsidy was to ensure that there would be enough honeybees for crop pollination. However, since most income came from honey and not from pollination, Congress subsidized honey production at prices that would enable beekeepers to maintain viable apiaries.

Historically, small farms with diversified elements often included bees; the 2.3 million bees of 1940 that I mentioned above were spread out on 260,000 farms. But by 1974, a similar number of hives, 2.4 million, were confined to just 6,500 farms and apiaries. With the consolidation of farms and specialization of crops, bees became more and more of a migrant labor force removed from the farm. The different labor specializations of the bee—honey production and pollination—became separated. This led to the weakening of the honey lobby, which could no longer say that by subsidizing honey the government could protect pollinators. The American Honey Producers Association and the American Beekeeper's Federation grew apart and now often hold their annual meetings separately; when they meet together, the conferences are divided into honey producers and pollinators.

Fifty years ago, honey prices could help sustain beekeepers, but now many beekeepers dump their honey or feed it back to their bees because almond honey is so bitter it is inedible. And while the honey from other crops is viable, it is still only marginally worth the effort and cost of extraction, processing, and marketing. So as the bee became more valuable as a pollinator than as a producer of honey or wax, the beekeeper became more tied to the agricultural industry. The beekeeper's relationship with the bees changed from keeping them healthy over time so they could cultivate a honey crop to transporting them so they could do the specialized labor of pollination.

Moreover, international market issues continue to haunt the honey industry. Currently more than half of the 410 million pounds of honey that the United States consumes is imported from China and Argentina (Woodruff 2012). And it remains the case that most US honey producers are sideliners or pollinators who augment their pollination business with honey production if the price gets high enough (Schneider 2011). Moreover, the value of bees in economic terms, which comes from a hive for pollination, is about 150 times that of honey and beeswax combined. To make matters worse, the decline is not only happening in economic terms; the production of pounds of honey per hive in the United States, once one of the highest in the world, has tanked (Hinyub 2012). As Jeff Anderson says, "The same number of bees just [does] not produce as much honey per hive; they are just not as healthy."[2]

So while some beekeepers still produce honey and ride the changing market, all industrial beekeepers rely primarily on pollination to remain in business. Many could not make the switch, so there are far fewer keepers today than at any time in the last century in the United States. There were also a lot more beekeepers in 1940, around 800,000, compared with the 100,000–200,000 that the USDA estimates today (USDA NRCS 2014). Possibly more significant in the long term is the average age of beekeepers, which is over sixty. Jeff says, shaking his head, "You walk into the annual meetings of the American Beekeeping Federation and you think, is this the beekeepers' meeting or an AARP meeting? . . . I'm not surprised with the state of things. I'm not sure I would recommend [beekeeping] to someone new."

In fact, there are new workers, but most of them are less visible than the older white beekeepers. While visiting bee- and queen-rearing yards, it is clear that there are young workers in the new beekeeping workforce, most of them Latinos. I have seen them driving the forklifts, managing the queen rearing, and building the hives. Not surprisingly, they are rarely part of the conversations and conventions on the future of the industry. I conducted interviews with around thirty Latino workers, mostly in Spanish, who work for the industry; only two of them aspired to run their own operations. Some pointed out that unless laws about driver's licenses change (as undocumented workers, they face deportation if caught driving without a license), they would not start running their own operations because of how dependent the operations are on mobility. For many of the others, beekeeping was just one of many seasonal jobs that paid well, and the initial cost to buy hives and vehicles prohibited them from getting into the business permanently. Julio Mateo stated, "I love working with bees, but getting started is just too expensive to make it work. You have to start with a lot of hives to make it pay."[3] The Immigration Reform and Control Act of 1986 has a special federal exemption that the American Beekeeping Federation lobbied for. It currently allows itinerant beekeeping operations and labor contractors to bring in workers from as far away as South Africa and Australia to work in the industry. These laborers still make up a very small percentage of the current workforce, and only a portion of them follow the bloom with the beekeepers. So at present, less than 5 percent of the beekeepers manage over 60 percent of the total beehives in the United States.

Jeff is in that 5 percent, and he says it is not easy keeping his business going. He patiently talks me through his current preparation for the bee migration, which begins every year in the dead of the freezing-cold Minnesota winter. Before he can begin trucking his bees, he has to bring their numbers up to the informal and variable industry standard of six to eight frames covered with bees out of a total of ten frames in a hive box. This assures that there will be enough bees to manage the acres of almonds that they will be tasked with pollinating. He starts by making and then feeding the bees a rich mixture of either corn syrup or corn protein patties (chemically laced power bars) with cheap corn oil and protein sources imported from overseas.

Steve Ellis, a friend of Jeff's who is also from Minnesota, joined us in our conversation at the breakfast table and then later at the diner. He told me, "This is a really strange time to start working on the bees, when there is snow on the ground. Bees should be balled up for a number of more months, but we've got to push them to bring their numbers up so we can start the season with strong, numerous colonies. So we put them on drip and keep them juiced up so that they are ready to go in January." That is when Steve and Jeff lay out their hives on pallets and load them with forklifts onto the back of semis headed for California. Many other beekeepers I interviewed rent tankers so they can mix their own corn syrup concoction and pump it into the hives each week to prepare the bees for the orchards.

Jeff and other beekeepers must drive their semis across the country, stopping at regular truck stops. Though the hives are tarped, thousands of bees usually manage to escape, which makes beekeepers, according to Jeff, "both very unpopular and very quickly served" at most stops. So keepers mostly move their bees at night, constantly worrying about the temperature and well-being of the bees, not to mention the possibility that the semi will flip over. They arrive and unload in large bee yards, where the hives are put on pallets and loaded onto smaller flatbeds that are driven through the fields and distributed at around two to four hives per acre, mile after mile. Jeff notes, "Putting the bees in the fields is tedious and grueling work, and it not as clean and simple as it sounds. The misuse of a forklift sends tens of thousands of bees into the air, the mud in the fields is brutal, you're constantly getting stuck, or you hit an irrigation pipe and it gets really messy. . . . There is always something, and you are doing this after having driven all night." This goes on for days at a time so the beekeepers can lay out the hives in the roughly 1,200 square miles of almonds planted with laser precision in California's Central Valley. He adds, "The fields are so precise that at times it feels like I am in an industrial maze rather than a fruit orchard." The timing is so important that beekeepers drive all night; to miss the bloom is to miss most of the year's income.

The industrial conditions have gotten so extreme and the cost and labor are so high that some keepers forgo these winter steps and buy new bees from Australia, where the season is ending, which are shipped by airfreight in thousand-pound batches. But as Jeff says, "If this fully happens, we are going to stop being beekeepers and become bee transporters. Just moving bees around without even trying to keep them through winters is not beekeeping anymore." Not only are beekeepers tired of the high cost of keeping bees over the winter, they are also tired of the radical increase in what is called "splitting." This is what beekeepers do to increase the number of their hives. They divide a large healthy bee colony into two separate smaller hives by moving frames of brood (bee larva), food pollen, honey, and a portion of the living bees to a new hive box where they then add or rear a new queen. One strong hive with a large population then becomes two hives with smaller populations; with the addition and acceptance of a new queen and time, labor, and resources, one hive becomes two separate marketable hives. Beekeepers split hives annually to deal with the losses from disease, accidents, pesticide kills, et cetera. Even when there is a significant loss every year, a small split hive with a healthy queen and a good nectar and pollen source can rejuvenate itself very quickly. All of the 120 commercial beekeepers I interviewed said the current level of hive splitting is a significant change in the industry. The older the beekeepers were (80 percent of those I interviewed have been keeping bees for more than twenty years), the more stark the transformation seemed to them. David Mendes, president of the American Beekeeping Federation, said that hive splitting is probably the most significant change in beekeeping in the last ten years: "When I started, we would split once or twice a year and we would have to replace 10–15 percent of our colonies to keep our numbers stable. But now most beekeepers will tell you they split over 100 percent—the total number of their colonies—every year, and if you ask them after a beer, they'll tell you the truth: that it is between 100 percent and 200 percent splits."[4] As David Hackenberg said, "It seems like the bees just want to die . . . and sometimes it feels like I hardly do anything else [but split hives]. . . . I don't feel like we are managing bees anymore—we are mostly just trying to keep them alive."[5] Steve and Jeff feel the same way and corroborate Mendes's estimated percentages in their operations, both admitting to "well over" 100 percent.

Many beekeepers told me that it might end up being cheaper for them to buy new bees every year than keep them over winter. Jeff admits that after one really bad year with high hive losses, he had to augment the number of his hives to meet his almond contract obligations: "I hated doing it, but otherwise I would have undermined people's trust in my ability to deliver on a good year. . . . You never know with beekeeping; you sometimes fool yourself into believing you are managing the bees, but most of the time you are at the mercy of the changing winds, be they mites, or CCD [colony collapse disorder], bad weather—so much [is] out of your control." He went on, "If you are going to hold on for the long haul, then you got to hold your breath sometimes, and sometimes you got to hold your nose to make it in the long run. I'm not proud of all that I've done with chemicals in the hive and corn syrup—none of us are. But none of us wants to give up keeping bees, and you cannot keep bees long term anymore without making short-term compromises." In fact, many beekeepers have gone out of business; it is estimated that one-third of the beekeepers in the United States quit beekeeping after the Varroa mite decimated the bee population and forced beekeepers to use chemicals and invest significantly in more labor to maintain hives. More have dropped out due to the rising costs of transportation.

After the almond crop is finished, beekeepers pack up and follow the bloom. Some go north to the apple and cherry crops that bloom in March. On the East Coast, they follow the citrus and then migrate up to the melons and blueberries. In the West, they go to Montana's alfalfa and clover fields at the end of the summer. This means that honeybees will eat only one type of food for a month and then switch to a different type, which, according to entomologist Gordon Wardell, has literally transformed the chemical makeup of the bee's intestinal tract.[6] In the fall, beekeepers head back to wherever their main bee yard is to clean up the hives, deal with the dead, and empty hive boxes. John Miller, a long-time beekeeper who coordinates twelve semis and two tankers of corn syrup, told me that last year he loaded his bees off and on each semi twenty-two times throughout the season.[7] Then everything starts all over again in December. He added, "We push the bees harder and harder every year. These are not the melancholy bees of poets and philosophers; these are fully industrialized worker bees, juiced on chemicals and miticides from Monsanto, fed high-fructose corn syrup and cheap soy protein imported from China, and genetically selected for survivability of trucking. These are stressed out, road-weary bees on steroids."

Jeff Anderson fills his second cup of coffee and tells me,

The cost in inputs just to keep the bees alive is a remarkable change. When I started over thirty-five years ago, we hardly, if ever, fed the bees in the winter. Maybe if a hive looked weak, but we just didn't need to. They came in healthy and there was enough forage in the fall that they didn't need it. . . . We never fed. But by 1990 we started [feeding them] a few gallons a year: two in the fall, two in the spring; now I am lucky if it is twelve gallons, three times that. And it is not just so they are robust enough to go into almonds in February, but just to get them through the winter. It used to be we would get our bees back to Minnesota after pollination, we would get a good dandelion flow and they would get healthy, but now there is a flow, and they seem to turn the other way or not be able to bulk up for winter at all. So if we don't feed them now, they won't make it through winter.

Steve, who lives alongside Jeff in a bee-yard trailer park with some other beekeepers during the almond harvest, adds, "The bees seem to be able to bring less and less [nectar and pollen] in, and they have a more voracious appetite." Jeff chimes back in, "Yes, yes, and yes."

There has been a concerted effort to increase bees' endurance and efficiency, most notably through what they eat. Honeybees make their own concoction called "bee bread," a mixture of gathered pollen flora nectar, enzymes, bacteria (over twenty-five), organisms, and fungi (almost two hundred kinds) that makes up the primary source of food for the hive (Schmehl et al. 2014). The resulting material is considerably more nutritious than unprocessed pollen but is susceptible to pesticides and fungicides. The extensive and intensive amounts of fungicides, particularly in almond orchards, have been shown to affect the microbes in bee bread, decreasing the ability of the bees to store and digest the protein that they gather. A healthy hive develops from the communities of microbes that underlie it. But monocrops, pesticides, changing landscapes, the use of high-fructose corn syrup, and new protein supplements in the bee's gut—a multispecies environment—have impaired the presence and vitality of the microbes, which in turn have transformed the bee's metabolism, digestion, and immunity. If fact, over 90 percent of the DNA in a bee's body, like the DNA in a human body, is not of the host bee. A complex interspecies interaction makes what we consider a bee a bee, and the radical changes to the dietary and chemical milieu in which a bee resides drastically change the biotic community that helps constitute a bee. The result has been significant and lasting effects on the chemistry and biology of the modern honeybee, and with them what it means to be a beekeeper (Gilliam 1997; Hoffman et al. 2009).

It also takes more effort and money to increase bee populations quickly and keep them alive and viable for pollination. With the changing diet and health of the bee, new products have entered the market, namely pollen patties. Most commercial beekeepers I spoke with either buy stuff with names like MegaBee, Ultra Bee, and Bee-Pro by the pallet-load or make their own by combining super high levels of protein with other nutrients into a thick mixture loaded with corn syrup and soy protein, dried egg yolk, citric acid (a preservative), vitamins, and yeast. Then they shape it into a patty that looks and actually tastes like a power bar. Protein patties were only invented in the past five to ten years, but now, Jeff says, they seem to be an absolute necessity.

The modern bee diet has been transformed not just by the industry of beekeeping but also within the industrial landscapes of urban beekeepers, which are the fastest growing group of beekeepers in the United States. Jeff comments, "People are always criticizing us in the industry. But have you seen the number of food and chemical products for the backyard beekeeper? Someone is buying that stuff. . . . I know that many of them don't use them, but if they lose their hives, they do not lose their business. Besides, few people think of the chemicals that bees are exposed to in cities." The most poignant example, and the one that both Jeff and Steve mention to me, took place in 2010 when rooftop urban beekeepers in New York started noticing that the honey in their hives was taking on a deep red color. The beekeepers, curious about the floral source of the red color, sent samples to a New York state apiculturist who found that the color was red dye no. 40, or in chemical terms, 6-hydroxy-5-[(2-methoxy-5-methyl-4-sulfophenyl)azo]-2-naphthalenesulfonic acid, an almost fluorescent red chemical dye. It turned out that a maraschino cherry factory located near the hives was leaking dyed-red corn syrup, which the bees had been actively collecting. This upset the sensibility of the beekeeper, who said in a *New York Times* article that "it feels like a betrayal to our sense of how nature should work. Shouldn't they know better?" (Dominus 2010).

The more consequential changes, however, have occurred in the diets of bees in the rural, industrialized agricultural environment. The quality and quantity of bee pasturage has been declining steadily over the past sixty years because of changing agricultural patterns. The rotation of cover crops has declined because of chemical fertilizers, the rise of feedlots has reduced key plants such as alfalfa because much of the nation's beef supplies now eat corn-based diets, and the harvesting of alfalfa before bloom, which maximizes the plant's protein content, has transformed the bees' diets. In many areas, the decreased ratio of open land to developed land has decreased productivity and, as a result, potentially affects bee mortality, sensitivity to disease, and chemical exposure.

Perhaps the most radical change in the agrarian landscape and the bee diet in the last fifty years is the ubiquity of farms that produce, as Jeff says, "soy and dirt, corn and dirt, trees and dirt. After the bloom there is nothing for the bees but dirt." With the advent of Roundup and especially Roundup Ready crops, the average amount of honey per hive has actually declined in the United States over the last thirty years. Jeff explains, "Most people don't notice the change, or kind of like it that the roadsides and sidewalks are clear of weeds, but if you are a beekeeper, you are ultimately dependent on them. Roundup has made the landscape and the farms a lot cleaner, but to a bee and a beekeeper, they are just green deserts."

Many of the beekeepers I interviewed agreed that Roundup and Roundup Ready crops have caused the most significant change in the honeybee diet. In the United States alone, a conservative estimate of 215 million pounds of Roundup are used to almost totally control weeds at farms, parks, schools, hospitals, roadsides, railroads, golf courses, and homes, and have thus radically transformed the landscape for the bee. Before, weeds and their pollen and nectar were a perennial part of every landscape, but the ubiquitous use of Roundup has changed the biotic world that the bees occupy as much as any other factor. Since 1974, glyphosate [N-(phosphonomethyl)glycine] has become the dominant herbicide worldwide. It is a highly effective, broad-spectrum herbicide with limited health and environmental effects (though this is more and more contested). Moreover, Monsanto, the company that manufactures Roundup, introduced and patented genetically modified seeds that are resistant to the chemical, so that a farmer can spray their fields and kill everything except the resistant crop. The result is what Jeff described: a landscape where there is nothing but the genetically modified plant and dirt.

What this means is that the landscape has quietly and radically changed over the last forty years; farms are now pollen and nectar deserts for most of the year, except when monocrops bloom simultaneously. The feast-or-famine landscape has made it impossible to keep bees full time on most industrial farms; they will simply starve. Forced migration, which used to be a means of boosting honey production, now has become a matter of survival for the nation's honeybees. So while the blooms in these large monocrops are remarkably beautiful to observe and are bountiful for the bees, the surrounding fields, with their tightly controlled weeds, become, as Jeff described them, green deserts. So bees, like farmworkers, no longer spend the year in one place, part of a complex relation of agrarian production, but have to migrate, as do their keepers.

There is, of course, an uneasiness in making a migrant labor comparison between beekeepers, immigrant laborers, and bees. Are bees simple commodities or do they produce the commodities and transform themselves? One element of this approach to materialism and the commodity is an understanding of Marx's methodology of historical materialism, which is perhaps the most prominent and nuanced way of theorizing the relationship between the subject and the commodity. His analysis is clearly consequential here; the transformation of the bee into a commodity has profound effects on the bee and its keepers. However, as important as this approach is, it also has its limitations. For in this formulation, bees and other commodities are still inert in history unless a human agent animates them, giving them the agency of their maker, not always predictably. Thus, the subject and objects of history become animate and inanimate in the story, and the concept of animacy is reduced to a dichotomy between those with consciousness and agency and those without. In fact, it is on the back of the bee that Marx makes the Socratic distinction separating the subject of politics from the object of politics. He writes, "What distinguishes the worst architect from the best of bees is that the architect builds the cell in his mind before he constructs it in wax" (Marx 1976, 284). Commodities take on an animated form, but we know little of the ways that the animals, objects, and elements have politics and constitutive effects without a narrow and humanist notion of consciousness—for example, how the grain of the wood resists the plane of the carpenter, the uneven consistency of yarn, the tenacious pull of clay in soil, and the uneven rates of germination of wheat. Or in the context of this chapter, how the life specificities of the bee's existence are consequentially linked to the ways in which the laboring subjects become who they are. How do the differential qualities of the material with which laboring subjects work (be it wood, stone, metal, plastic, or flesh), not just the structural conditions under which they work, matter in the making of workers, apian and human?

Jeff and Steve are among a group of beekeepers who fervently believe that at the heart of the conditions in which they operate, a new class of systemic pesticides is significantly affecting honeybee health. What is significant here is the change in pesticide exposure. Jeff tells me, "It used to be that you could move your hive if you knew when someone was going to spray. You waited, and the effects on your bees, if they did not get hit during the spraying, was minimal." Steve adds, "Now with systemic chemicals that are in the plant, bees are constantly getting exposed to the chemicals through nectar and pollen from the plants. They never are free of the toxins." In this case, the toxins are well hidden, and while it may look like they are less toxic, many researchers and beekeepers believe the cumulative effects are significant. If the diet makes the animal, then the modern bee is a very different animal than it was even forty years ago.

But the changes in the conditions of the bee are more difficult to assess now because of the decline of an active, educated group of extension agents that used to be part of the agrarian landscape. While it is a less-discussed part of the changes in beekeeping, the transformation of state agricultural extension agents is dramatic. These officers used to monitor and help beekeepers by investigating pesticide spray incidents, monitoring and treating diseases, limiting the movement of diseased hives, and inspecting the health of bees being transported. While they could and did at times police and even fine beekeepers for the improper use of chemicals and destroy diseased hives, they also served a vital role for information exchange and as advocates against pesticide companies and the farm lobby. After hearing references to the declining extension programs, I asked my research assistant to call all forty-eight states in the lower United States. It turns out that over 80 percent have entirely lost their state apiary inspectors since the year 2000. In the remaining 20 percent, most state apiary inspectors have duties beyond beekeeping to attend to and receive less support and money for inspections. Moreover, I found that they receive almost no money for research, and in many cases agents do not have vehicles or money to perform any field inspections. People who used to be some of the most stalwart advocates, able to spot bee diseases and pesticide kills and give advice outside of the profit-driven industry, were fired just when some of the greatest challenges beekeepers have ever faced arose. It also meant that a coordinated effort to deal in a systematic way with the declining bee population and the precarious position of beekeepers became less possible. As Jeff says about the Minnesota inspectors, "We lost ours a few years back. As state money was cut, they asked beekeepers to pay for it, but who is going to pay to be regulated? They could not come up with the funds so we have no one anymore. . . . Looking back, it probably would have been worth the funds."

Steve gives the specific example of growing resistance to a chemical Varroa mite treatment: "We would have been forced kicking and screaming, but collectively we could have managed the treatments in less of an individual, self-interested way; that would have ultimately avoided the speed and extent of the mite resistance to the treatments." As a result of the decline in apiary inspectors, there is almost no oversight and little state-level advocacy for beekeepers by educated and informed professionals outside the chemical companies and product distributors who visit beekeepers. The modern bee maladies have as much to do with economic privatization and austerity as they do with pesticides and pathogens.

The way industrial honey production has reorganized the labor of the bee is profound. The bee's role in pollinating in the context of industrial agriculture has transformed bees, beekeeping, and the political-economic geography of beekeeping itself. Virtually all commercial beekeepers in the United States must now move their hives to stay in business. Jeff admits, "There is something fundamentally wrong with beekeeping today. I am part of that problem, but if I change, I am out of work. I need to cut corners; in fact, I am rewarded for my beekeeping vices. I cannot say I do not have a choice." He smiles at me and adds, "Don't ever forget, beekeepers have choices and the choices are clear and ours. We can go to almonds, we can split our hives, we can put poisons in them to kill the mites, or we can go out of business. . . . It's all a very fragile balance." He looks at Steve, who sits back in his chair and looks out the window of the double-wide. Jeff looks down at the very bottom of his empty coffee cup, rolls it on its edge, and drinks the last drop, changing the subject to the upcoming cherry bloom in Washington.

There is, finally, a deep irony to the decline of bees. After feeding them corn syrup and pollen patties, importing and splitting their hives, and doing just about everything humanly possible to increase the number of bees for the almond crop, at the end of the season, beekeepers often must bring their numbers down to truck the bees back to their winter storage sites. This means killing the bees by dumping or poisoning them. Killing bees is a strange but necessary part of industrial beekeeping. To do anything else would be, as one long-time beekeeper told me, "economically wasteful." He is right, and he hates killing off massive quantities of bees, "but it is simply market logic." He says if he does not kill off bees before wintering, his colonies will swarm, and he will lose healthy queens; he can only truck, store, and feed a certain number of bees over the next winter before the process starts again. While the dramatic losses due to CCD have reduced the practice of killing bees, because the bees reduce themselves, it is still a stark economical violence that is a necessary part of contemporary industrial agrarian life.

Bees, keepers, chemicals, bacteria, semis, tariffs, pollen, and laborers have constitutive histories within this industrialized landscape that they inhabit and make together. Marx sought to demonstrate that the naturalized binary between subject and object made the commodity appear as if it were simply an inert thing. What he did brilliantly in the first section of his great work *Das Kapital* was demonstrate that the commodity is not inert but animated by the politics of its production. Commodities are infused with congealed labor that is invested in them, determining, in part, their value and the form they grow to take. For Marx, the fact that commodities are perceived as having an intrinsic value is in fact a fetish that depoliticizes their making and the politics invested in their form. The process of making them commodities also erases the particularities; they become equivalent and exchangeable through the congealed labor, erasing the specificities of their forms and eventually forming them in the shape, literally, of what they are most valued for in the domain of exchange (Marx 1976). Here, diverse forms become congealed in what we have come to know as The Bee and The Keeper, and they come to live in conditions that make them into a form that defines them. For the bee, both the normative biological taxonomy itself and the material species form are produced through the economic relations of production; the bee becomes worker, remaking itself and its socialities through its relationship to work. This form of The Bee becomes the starting point of much of our research into bees, rather than a product of already complex and political histories. This ultimately conceals the animacy and particularities of commodities, making the complexity and vitality of bees into a marketable abstract form that is produced as long as it is profitable, and becomes disposable as it becomes unprofitable. Moreover, the bee becomes an object, the biological form of a commodity that assures its existence; the consequences of letting the bee die are too high, but this role as an agrarian worker also produces the vulnerabilities that bees now suffer.

Bees become the commodities they are through what makes them valuable, primarily as pollinators, and become disposable, along with multitudes of other commodities, in a system that encourages consumers to buy in ever-increasing quantities and dispose of things in ever-shortened time frames. This, in part, made both the bee and its keeper. As valuable as this analysis proves to be, there are limits to this approach, for the animacy of the bee here is largely a function of the way it is produced as a commodity; this is still an indirect form of human agency, just transposed onto the bee as commodity. How might objects be more than commodities formed by social relations? How might objects be formative of both bee and human socialities?

But perhaps the clearest demonstration of bees' labor is that they have become abstract objects of labor relations in ways that have defined their biopolitical vulnerability and essential labor at once. But work here is clearly more complicated. It is not just a reconsideration for the agency of bees but that the conditions of industrial agrarian labor transform and are transformed by the working relationship of more than human relationships. The very possibility of much of industrial farming is made possible by bees' annual migration on semitrucks; and their labor relationships are constitutive of the very form and size of industrial monoculture farms. At the same time, the contracting of their labor (that mediates and profits from their role in agrarian production and the subsidizing of cheap fruit and vegetables with their bodies), social relationships, the structure and form of their living conditions: these in turn produce both surplus value and tattered bodies, and they are all signs of the modern conditions and constitutions of bees as workers. The essential and socially transformative work that bees do and the precarious conditions under which they live make bees workers, and to miss this fact is to miss the political possibilities of a broader and deeper critique of capitalism.

Notes

1. All Steve Ellis quotes, personal communication, Oakdale, California, March 11, 2012.

2. All Jeff Anderson quotes, personal communication, Oakdale, California, March 11–12, 2012.

3. Julio Mateo, personal communication, Bakersfield, California, February 3, 2016.

4. David Mendes, personal communication, Anaheim, California, January 7, 2015.

5. David Hackenberg, personal communication, Lewisburg, Pennsylvania, November 12, 2011.

6. Gordon Wardell, personal communication, Baton Rouge, Louisiana, January 10, 2014.

7. John Miller, personal communication, Anaheim, California, January 8, 2015.

Part Three

Futures of Work

CHAPTER NINE

Cultural Analysis of Microbial Worlds

JOHN HARTIGAN JR.

This is an account of bacterial flagellar motors and the social labor they perform. Since Karl Marx, labor studies has asserted that the social character of labor is distinctive to humans, in contrast to "natural" activities of nonhumans. Flagellar motors, though, produce their social labor in specific cultures of agar cultivated on petri dishes. Building off this cultural basis for bacterial social activity, I take a recursive approach here, articulating a series of iterations of key terms—"motor," "labor," and "culture"—scaling their connotations between the two poles, human and bacterial. My aim is to scale down social analysis (as we know it with humans) to apply it to bacteria, while scaling up biologists' use of culture so it models a form of analysis applicable by cultural anthropologists. The basic technique here involves recursively orienting the core terms to a different set of connotations by loosening their association with a human referent, creating a feedback loop through which the results can be brought back to transform the starting point of cultural analysis assumed by anthropologists. This entails loosening both culture and labor from an equation with what makes humans unique; disrupting the parallel association of nonhumans with nature allows us, then, to recognize manifold forms of sociality and laboring collectives beyond the human.

Motor 1 (Mechanical)

The bacterial flagellum is driven by a bidirectional rotary motor, which propels bacteria to swim through liquids or swarm over surfaces.
—Jonathan D. Partridge, Vincent Nieto, and Rasika M. Harshey, "A New Player at the Flagellar Motor: FliL Controls Both Motor Output and Bias"

Flagellar motors are a major topic of interest for understanding bacteria, life forms that make up about half of the cells in the human body. If we regard this usage of "motor" metaphorically, we would miss an opportunity to rethink theories of labor in a manner that more broadly encompasses aspects of how nature works. In terms of basic physics, it generates power through a combination of torque, angular velocity, and rotational speed. Since it routinely outperforms motors designed by humans—rotating five times faster than a Formula 1 engine, propelling bacteria dozens of body lengths per second—perhaps our ideas of motors are rather limited, if we principally associate them with human contraptions. But more than individual motors, what is really interesting is that when they work together, social swarming occurs in bacteria.

Bacterial swarming fascinates Rasika M. Harshey, a professor in the Department of Molecular Biosciences at the University of Texas. We spoke together in her office shortly after she was awarded a five-year research grant from NIH to study motor sensing in bacteria. Broadly, bacteria use the rotation of helical flagella to propel themselves through liquids or across a solid surface if a liquid film is present. Their bidirectional motors rotate clockwise or counterclockwise in response to chemical signals regarding the environment and its potentially favorable aspects. This is well established. Rasika's research focuses on how the motor shifts from being at the output end of sensory detection to itself perceiving some type of surface signal about its environmental locus. Responding to such signals, bacteria can decide to grow more flagella and swarm over the surface or make themselves into sedentary biofilms.

Swimming is what bacteria do naturally in an aqueous medium: an individual cell's motion is achieved by rotating flagellar filaments. In order to swarm—to move in concert over a hard surface—bacteria resolve several mechanical challenges by attracting water to lubricate the surface, overcoming frictional forces and reducing surface tension. I asked Rasika about the swarming behavior I had watched in videos that she provided as supplemental materials to the article in *Journal of Bacteriology* she coauthored with her postdoc, Jonathan Partridge. She said, "They show the swarm, and it's advancing on a surface. And you watch the bacteria moving and see them creating these whirls and swirls."[1] The social patterns are mesmerizing and frantic, but how to describe them? "So you can see that there are these . . ."—pausing unexpectedly before she spoke the term they use in quotes in their article—"*packs* that go together. But they're moving in all kinds of directions." I asked about the surface in question—various concentrations of agar—and the curious matter that this behavior has almost exclusively been observed only in laboratory settings. What technology did they rely on to accomplish this feat? "Well, you know," she said, laughing, "it's just the age-old petri dish!"

Culture 1 (Bacterial)

Biologists use "culture" alternately as a verb or noun: as something they do to make a life form grow or to identify the medium in which it propagates. My colleagues in the life sciences at the University of Texas offered me multiple, nuanced elaborations of how they use this term, but they all gravitate around these two grammatical forms. For a more standardized view, I talked with Marsha Lewis, who developed and teaches the curriculum for the Bioprospecting Research Stream, a large-enrollment, inquiry-based course that is part of UT's Freshman Research Initiative. She told me that, for the students, "learning to culture is like a rite of passage. Because it's basic to everything we do."[2]

Marsha added that this is also because there are so many ways to concoct it, and so many ways its many variations can produce unintended results. "Culture changes if you use a different type of water. It might set off a whole different chain of metabolic pathways even if you just change the water grade." The students don't realize how tricky this can be, even though they have so much in common with the medium they're formulating. Marsha observes, "It's just like with people: if you put them in a different environment, they're going to have a different outcome."

Creating culture involves nutrients, in a broth, and a material foundation, such as agar—a gelatinous substance (polysaccharide agarose) derived from seaweed by boiling it until the supporting structures of its cell walls dissolve. These concoctions can be enormously varied and inventive, Marsha told me,

> That's where I kinda let them loose and have fun. They can make a V8 Juice–based culture. Or just the standard, LB [Lennox broth], store-bought media. But I encourage them to get creative. So if your fungus likes a leaf from a magnolia tree, why don't you take some magnolia leaves, boil them so it's aseptic, there's nothing alive in there, then autoclave it, but basically make a tea. And y'know, hopefully you don't destroy some of the good stuff. Then you make a plate, as an agar, make a plate, and maybe you can *entice* some fungi to grow, because the objective is also . . . it would be interesting to isolate new cultures. So the difficulty is, there's all these microorganisms, but not all of 'em are *culturable*.

"Entice" caught my attention because it entails persuasion perhaps to the point of seduction—lured in with nutrients. But her last point is quite striking and important to understand: only about 1 percent of bacteria are capable of being cultured in this manner. This is quite significant for how biologists use culture to pose and answer questions of bacteria.

Labor 1 (Work)

A motor turns energy—heat, electrical, pneumatic, or chemical—into mechanical output to perform work. Flagellar motors are powered by a flow of protons generated through an electrochemical potential imbalance across a cytoplasmic membrane. This output is distinct across bacterial genera: *Salmonella*'s motor spins at about 300 Hz, while *Vibrio*'s is much faster, at 1700 Hz. But they all share a similar structure: a basal body composed of a series of rings, the hook or universal joint, and filaments or helical propellers, plus a number of stators (the stationary part of a rotary system); torque is generated by sequential rotor-stator interactions coupled with ion flow that is "channeled into the mechanical works required for flagellar motor rotation" (Minamino, Imada, and Namba 2008). The bacterial flagellum is a rare example of a biological "wheel," a basic tool of transport crucial to human society. Most remarkably, these motors self-assemble; their forty or so components build themselves in a linear process, working out from the base, forming the hook and then the filaments—the genes with instructions for building the requisite proteins for all of this are clustered as operons that activate and deactivate in sequence through the assembly process.

The output of each motor produces two distinct motions: counterclockwise rotation makes the bacterial cell "run"; clockwise rotation makes it "tumble." They are also distinguished in that running seems directed, while tumbling is characterized as erratic or random. The terms for these motions are consistently used in quotes, in the literature and when Rasika and Jonathan talk about what they are observing. While such usage is not fetishizing, in the sense of projecting animate capacities onto inanimate objects,[3] they do seem to hesitate over the gap between such animal motions and what is occurring at the bacterial level.

They have no hesitancy with "swimming," since the self-propulsion of a body through liquid by a flagellar appendage seems unproblematic; indeed, bacteria's rapidity with this movement—up to sixty body or cell lengths per second—far outstrips the speed of human swimmers. However, the use of "flagella"—Latin for "whip," perhaps the single most emblematic technology of forced labor—links far more utilitarian connotations than the athletic or recreational associations of swimming suggest. "Run" and "tumble" are also used with circumspection because they pertain to apparent search strategies bacteria employ in locating food or favorable eco-niches. These motions are checked or harnessed by the collective motion of swarming.

Motor 2 (Swarming)

What is the purpose of bacterial swarming? Rasika and Jonathan suggest it is colonization: acquiring territory and increasing population size, which involves many challenges and obstacles. "Bacteria overcome these obstacles by banding together in large numbers, exhibiting a plethora of mechanisms to attract water, to increase flagellar motor power, to lubricate the cell-surface interface, and to reduce surface tension" (Partridge and Harshey 2013, 910). This plethora of mechanisms is generated in the course of species' adaptations to distinctive niches, which is a far larger subject than this chapter can treat. I am mainly interested here in how sociality increases motor power. Individual bacterium swim, but side-by-side cell movement results in swarms, as the number of flagella increase on the surface of individual cells. Additionally, these bacteria secrete surfactants, amphipathic molecules that reduce tension between the substrate and the bacterial cell, thus permitting surface spreading.

In a swarm, the mechanics of bacterial movement change. The same flagella that were inadequate for moving across surfaces now are functionally sufficient. Some coordination between cells occurs, permitting unidirectional movement, which even allows for reversals of direction through group propulsion. Whether this behavior is the result of developmental changes remains an open question. But the alterations seem mostly to be metabolic. The important point is that, as a collective, the bacteria are able to accomplish work. "Cells travel in a group, likely because they collectively *trap* a larger pool of water, because they *entrain* each other, and/or because the *larger torque produced* by a group of cells might be more effective in breaking the surface tension of liquid for forward motion" (Partridge and Harshey 2013, 915; emphasis added). The possibility that surface contact is the direct stimulus impels Rasika and Jonathan's research on motor sensing. They hope to understand these group dynamics via increasing capacities to observe individuals within a swarm. My questions are more directed to what is occurring socially with the bacteria, how individuals are interacting; how are they interpreting local contexts?

Culture 2

I met Jonathan Partridge in the lab, where he made time to talk with me about their research on swarming, past, present, and future. I told him about my interest in culture, and he somewhat bemusedly showed me their agar. First, the barrel of "the normal agar. This is what most people buy, that's regular"—the Difco brand.[4] But they prefer working with Eiken agar, because it is much finer. Then he tells me a story.

Early in her career, in the early 1990s, Rasika was mixing up some agar plates and made a mistake. Instead of the typical concoction at 1.5 percent concentration, she mixed the agar at 1 percent, creating a softer surface. The bacteria she was working with at the time, *Serratia marcescens*, behaved bizarrely in this culture, Jonathan said. "Instead of them just being plated [fixed in place], they were swirling out. They were doing this," he said, tracing out swarming motions with his hands. Also, they had a pigmentation to them, instead of being clear. Perplexed, Rasika tried to understand what was happening, searching the literature, querying other researchers, and isolating mutants that could not swarm. A Japanese scientist, Tohey Matsuyama, was interested in these mutants to test if they were missing a powerful surfactant that *Serratia marcescens* makes. He reported back that some of them indeed were not making the surfactant, but that all of the non-swarming mutants he tested were perfectly capable of swarming. Through this exchange, Rasika realized that the Japanese agar (Eiken agar) was facilitating the swarming. This agar is somehow "wetter"; it holds more water on its surface. "So the understanding in the community now is that a moist surface is critical for swarming to occur, and this can make a huge difference to how reproducible the results are," Jonathan told me.

By tweaking concentrations of this culturing medium, Rasika and Jonathan deduced that there are two different types of swarmers. Robust swarmers can navigate the surface of media with a higher concentration of agar; members of this group include *Azospirillum*, *Proteus*, *Rhodospirillum*, and *Vibrio* species. In contrast, temperate swarmers can only move on softer agar surfaces (0.5–0.8 percent agar); members of this group include species of *Bacillus*, *Pseudomonas*, *Rhizobium*, *Salmonella*, *Serratia*, *Yersinia*, and *E. coli*. These arrangements of cultures and species point to distinct mechanics for overcoming frictional forces. "Robust swarmers synthesize many more flagella than temperate swarmers," suggesting their function is to provide more thrust for crossing harder surfaces (Harshey and Partridge 2013, 911). As well, robust swarmers substantially increase in cell length (>30 μm), providing more surface area for more flagella, thus increasing stride and minimizing drag. Temperate swarmers do not show these changes—or do so only in a more limited manner—relying, instead, on special stators or stator-associated proteins. This indicates the importance of a set of stators "that can work more effectively against friction and must transmit more power to the motor," Jonathan said. Across such a range of species it is difficult to formulate one uniform account of how swarming works among genera of bacteria, but the contours of the dynamic are clear: "Multiple solutions work to increase thrust," he intoned. Culture—as medium, as collective motion, and as lab work—is crucial to the analysis of the mechanics at work in swarming. Rasika and Jonathan adjust the medium to create swarming, because, again, it is not clear that this occurs in nature.

Labor 2 (Transporting Cargo)

Such behavior may confer a means of long-range communication, such as circulation of nutrients encountered by the advancing edge of the swarm to inner or older regions of the colony, transport of signaling molecules, more efficient acquisition of oxygen from the atmosphere to reach bacteria under a multilayered pile, and better regulation of temperature.
—Jonathan D. Partridge and Rasika M. Harshey,
"Swarming: Flexible Roaming Plans"

Rasika and Jonathan speculated more broadly on these dynamics in a review article, "Shelter in a Swarm," in the *Journal of Molecular Biology* (2015). One aspect of swarming they highlight is the bacterial capacity for transportation, or "carrying beneficial cargo." Again, there is an option to regard this usage of "cargo" metaphorically, insisting on the primacy of the human referent; or we can opt to ask how the labor of transportation in these colonies opens up a perspective onto forms of sociality in bacterial communities. The latter option leads to the division of labor, the foundational subject for social analysis, as articulated through Karl Marx, Herbert Spencer, and Emile Durkheim, but staged now in nonhuman realms. Here is how Rasika and Jonathan formulate the matter: "A common theme that appears to be emerging in bacterial communes, whether in swarms, during flagella-independent migration, or within biofilms, is that bi-stability of certain traits can set up a phenotypic heterogeneity that leads not only to bet-hedging as has been previously proposed but also to division of labor" (Harshey and Partridge 2015, 3686). The type of phenotypic changes noted with swarming (increased cell length and growing more flagella) becomes the basis for phenotypic heterogeneity within a swarm. That is, certain cells in an isogenic population—sharing a homogenous genetic composition—physically transform in response to complex signaling, such as quorum sensing. Nongenetic variation (collective signaling) induces a bacterial population to bifurcate into distinct, coexisting cell types, which then play distinct roles in maintaining a community. This phenomenon—broadly observed, in and out of the lab—matches that condition that Durkheim (1997) used to contrast "mechanical solidarity" (homogenous) with the differentiated components of "organic solidarity" that characterized the division of labor. This concept was key for Durkheim in regarding differentiated individuals as fonts for upwelling of social solidarity and the capacity for self-sacrifice.

This differentiation occurs most pronouncedly in motile species but emerges in nonmotile ones as well. *Paenibacillus vortex* swarms, they note, have been reported to transport fungal spores and even other bacteria as cargo over long distances, but this activity requires multilayered swarms. Perhaps this is yet another instance of mutual cooperation in nature, even between bacteria and eukaryotic organisms. But the transport of cargo might also frame the possibility that phenotypic heterogeneity results in a division of labor: "For example, swarms of *P. vortex* appear to have two phenotypic variants, one more motile than the other; the hypermotile population spearheads colony expansion, has lower ATP levels, has higher tolerance to antibiotics, and has cargo-carrying capacity" (Harshey and Partridge 2015, 3686). But nonmotile organisms can function this way, as in a colony of *B. subtilis* that expands by differentiating into two kinds of cells, one producing extracellular matrix and the other one producing a surfactant: "Matrix producers form bundles that are hypothesized to drive outward expansion of the colony, which is facilitated by the surfactant producers" (3686). Rasika and Jonathan conclude from these examples that "the phenotypic heterogeneities observed in both motile and non-motile bacterial colonies are changing our notion of bacterial communes, where a deceptively uniform colony is quite non-uniform in reality" (3686). These communes now appear to be socially diversified but at the expense of developing a division of labor around the transport of cargo.

This concept of cargo is articulated across an array of microbes and is theorized as much in terms of fungi and protists as bacteria. A review article by one of Rasika's collaborators, Gil Ariel, focuses on microorganisms that facilitate the dispersal of other species in mutually beneficial multispecies swarms (Ben-Jacob et al. 2016). While bacteria can be transported by slime molds and amoebas for grazing or farming,[5] they highlight the division of labor that occurs in instances of bacteria conducting other bacteria. When *Paenibacillus vortex* swarm—producing highly motile "explorer cells" on the fringes and less motile "builder cells" in the interior (phenotypically distinct)—they can carry other bacteria, which are "considered cargo if they are moved but do not contribute to the motility of the swarm" (262–63). This species can move objects far larger than themselves, such as fungal spores, and "the size disparity between cargo and transporter and the distance covered challenges the way we think about scales in microbiology" (266). The matter of scale here includes notions of infrastructure (mycelial bridges, fungal highways, and traffic streams of sessile species "that actively promote their own dispersal by stimulating the motility of nearby bacteria"); forms of "traffic structure" are modeled via "culture on low-solidity agarose using tracking beads to aid visualization," which "revealed long-term stable traffic of motile bacteria" (264). More on infrastructure and beads below.

Motor 3 (Lab Work)

In trying to figure out how flagellar motors work, Jonathan examines the role of a highly conserved protein, FliL, which is found across all bacterial genera. In describing its various functions, he first explained to me the "regulatory cascade" by which the flagella assembles itself through three stages, from which emerges a secretory apparatus, then the bushings, and finally the hook. As each stage is completed, a signal initiates the next one. FliL plays a fundamental role in the motor operation, where it "has two separable functions, one providing increased motor torque and the other stabilizing the rod." Along with FliL's various protein/protein interactions, "it's doing something with motor output." Jonathan explained the innovative and precise lab work required to understand this output, saying, "The old way of doing it was we would shear the flagella off, which just left the hook"—the bendy, flexible region just outside the cell, attaching to the filament. They would affix the hook to a microscope slide and then watch the body as it turned on the hook. "Like this," he said, showing me with his hand rotating around the axis of an extended forefinger. Then they would count its rotations and how often it switched directions, counterclockwise and clockwise, running or tumbling. The problem, though, was that the entire weight of the cell body rested on the hook, affecting its rotational capacity, making it about fifty times slower. "You could still recognize patterns and the changes," he allowed, but the motor was doing the wrong kind of work to gauge its function in swarming.

They are working to adopt a new system for measuring its output. As Jonathan explained, "Now we stick on the whole cell body to the slide," after passing overnight cultures of *Salmonella* and *E. coli* through syringes to shear off the flagella. He said, "Then we have microscopic polystyrene beads that affix to the end and approximate the load or the weight of the full filament." They observe the rotational motion of the beads via phase-contrast microscopy, plotting their motion and directional changes. Jonathan shows me the graphs on his laptop from his recent presentation of this research at a conference that summer, "Collective Dynamics in Microorganisms and Cellular Systems." In the old measuring system, motor output was reduced to about 5 Hz; with the new method, *E. coli* motors rotate at about 200 Hz, "so it's much more accurate," he noted. When the FliL protein is knocked out through mutations, the rotation is slower and there are fewer changes in directions, so it affects both swimming and swarming.

To measure the motor's torque, they increase its load, replicating high loads by increasing viscosity of the liquid the cell is in, using the agent Ficoll. Altering the concentrations of Ficoll produced different degrees of drag on the motor's angular velocity. Torque—force required to spin the motor—increased or decreased with differing levels of viscosity, allowing them to compare wild type cells with mutants lacking FliL. The wild type motors performed consistently while the mutant ones produced less torque with higher loads. This finding underscores the important role of this protein in swarming over a hard surface, which equates to a high load. But it also presents an instance of "work" doubling as a subject of analysis, at the level of the lab and at bacterial level. What form of cultural analysis does this require?

Culture 3

I was most excited to talk with Rasika about their work with a border-crossing assay, one in which *Salmonella* strains are "allowed to migrate into an antibiotic-containing chamber" (Harshey and Partridge 2015, 3685). It is a topic that excites her, too, because it involves the remarkable capacity of the social bacterial swarm to overcome antibiotics. She sketched both the broader interests in this demonstration and the increased perplexity it induces, explaining that what interested people about the "Shelter in a Swarm" article is that "we talked about one property of a collective—its ability to withstand more antimicrobials. But we don't understand that at all. I don't know if this kind of a motion has anything to do with it, at all. Or if this kind of motion is generating some kind of structure in the colony that's helping it. We just don't understand how they're overcoming it."

Rasika tells me how they approached the problem, saying, "We set up this border crossing assay," then laughs boisterously. Acknowledging this odd outburst, I suggest that there are "loaded" meanings or "freighted" connotations to "border crossing" in Texas—invoking the labor metaphors common in the kind of cultural analysis I practice. She agreed but laid emphasis on how this research might provide "models to derive insight into the swarming behavior of large animals such as schooling fish or flocking birds" (Harshey and Partridge 2015, 3685). The social character of their labor in the lab, I anticipated, was ideological; the social character of the work of swarming, Rasika suggested, crosses manifold species lines. Migration is an active dimension of colony life for bacteria, a function of the specialized tasks resulting from division of labor among cells and a component of their emergent collective properties. Why should I assume these dynamics are only a "screen" or displacement of human political concerns in our state of residence? But what of borders and boundaries more generally?

"So again you've got that petri dish!" Rasika said with gusto, perhaps appreciating how simple and direct this form of basic science can be. "We sort of divided it, so there's a border. And we poured it with this solid media, and on one side there was no antibiotics and on the other side there was. So we started them here," she said, using her hands to delineate one side of the petri dish, "and they all got to the border and they had to cross over." The first stage of the experiment was performed with individual swimmers: "This is not when they're doing the collective thing," on the harder agar surface, she explained. "They are just swimming individually. They are not a collective." The results were to be expected. "When the swimmers come to the border, they are just stopped dead by the antibodies," she said. But the shock is that, in a later staging of culture, "the swarmers come to the border; they go right across." It is elegant in its clarity and simplicity. All they did was change the culture.

This is also an example she turns to because it impacts her thinking and sets the stage for the larger question, the one about sensing and acting as a collective.

"It's not like they're totally resistant; they're dying, in the population. But the group, as a whole, is surviving," Rasika said. I ask her what is promoting this survival. "So, some of these guys are sacrificing themselves," she said, punctuated with another nervous laugh, but this one acknowledging that "sacrifice" is a technical term in evolutionary theory and the academic division of labor is quite clear. "I use this word, but I'm sure evolutionary biologists have a very strict meaning about what that is."[6] Still, she opts to go there: "Let's talk about altruism, so the group survives," she says, and I cheer her on. But she cautions, "We're just nowhere near understanding this at all. We're just trying to figure out what we can understand."

Labor 3, Ideological

"Some problems can be solved only when individuals act together. This applies to bacteria in the same way that it applies to humans." These opening lines from the authors' summary of the article "From Cell Differentiation to Cell Collectives: *Bacillus subtilis* Uses Division of Labor to Migrate" (Van Gestel, Vlamakis, and Kolter 2015) could be considered a most general gesture at attracting the widest range of readers. But they highlight the fundamental concern in working analytically across species lines. In aiming to go "beyond the human" (Kohn 2013), cultural anthropologists "traffic" in a series of parallels and doubling referents and connotations that have often served as fodder for naturalizing social relations of dominance and subordination. In one regard, there is a fundamental contrast with such parallels, at least with racializing assertions and perceptions. Instead of drawing a more limited line around the human by dehumanizing through projections of simianization or as pests (Raffles 2010), "in the same way that it applies to humans" invokes broad forms of sameness (in contrast to otherness), linking humanity as a whole with nonhumans—Charles Darwin's basic gesture. Where a focus on metaphor fixates on the "leaps" across different domains, underlying dimensions of similarity between humans and nonhumans suggest a means of scaling social analysis "up" and "down" multiple levels of life forms.

To the extent we share some genetics with all life on this planet—and because our course of evolution has so many commonalities with other species, especially around the development of culture and sociality—these forms of sameness need to be the focus of analysis, even with the risks entailed by centuries of work at othering fellow human beings. Such an approach still needs to critique ideological operations naturalizing market forms, from cargo to the highways upon which they are transported; but the focus on sameness does suggest as much potential as peril in going "beyond the human." Indeed, the very capacity to engage nonhumans in the type of refashioned anthropology imagined by Eduardo Kohn may require letting go of an insistence that our key terms are only and always metaphoric when applied to nonhumans or that association with humans should be considered their real, "concrete" meanings.

It may be disturbing to see migration (of bacteria) talked about this way—in Rasika's assay or in the title of Van Gestel, Vlamakis, and Kolter's article on division of labor—given the high political stakes of this dynamic among humans. But Darwin used the term in exactly this manner, along with "community," long before these became the subject of social analysis as we apply it to humans today. Migration is, of course, practiced by many species; likely, the Americas were populated by humans following migrating animals thousands of years ago. Community, too, as invoked to characterize bacterial collectives, is widely evident in nonhumans—the breadth of social species across taxa is broadly acknowledged in the natural sciences. We cultural anthropologists may want to reserve the human referent here as primary, or insist, as Frederich Engels did with "competition," that Darwin's usage of these terms are reflections of his society and social position. But such gestures, in leading away from understanding forms of sociality among nonhumans, only double down on *our own ideology of an autonomous sphere of culture*, one that we inherit from the work of Alfred Kroeber. This ideology sanctions an ignorance of biological processes and animates a much older form, pollution ideology, one by which traces of the biological can be regarded as contaminating the purity of cultural analysis.

Analysis

What is happening culturally with these bacteria? This question scales in at least a couple of ways. First, they are being cultured to swarm, to produce a behavior that remains to be identified in natural conditions. Secondly, in these swarms, their motors are producing social labor, a result of some form of cultural activity that produces a collective dynamic. As a cultural anthropologist, what are my options for analyzing such labors as an ethnographic subject in these settings? With the first, my previous orientation would be limited to declaring these facts about swarms are social constructs, products of ideological labor. But that approach hobbles my ability to move cultural analysis beyond the human. To get at the second—and the potentially interpretive work of the bacteria in constituting a social collective—I need to frame the bacteria and the social character of their labor as the subjects of cultural analysis. Though bacteria are increasingly fodder for theorizing by cultural anthropologists, can they be rendered as ethnographic subjects as well?[7]

This question begins with Stefan Helmreich's *Alien Ocean: Anthropological Voyages in Microbial Seas* (2009), an ethnography of oceanography focused on microbes. In this book, "life itself" becomes a subject of direct address. Helmreich recounts, "If life was once believed to reside firmly in the territory of nature, . . . a riptide in the shifting currents of nature and culture" results in "a dissolving binary I have sought to mark in this book by using the suggestively overlapping terms of life forms and forms of life" (2009, 9). His study of oceanography offers a novel stance on anthropological theorizing—"working athwart," where the ethnographer is positioned alongside marine biologists—but it draws up short in positing or even speculating on an ethnographic means of engaging microbial life forms, rather than their representations in laboratory or field settings. Microbes may provide novel material for the intellectual labor of cultural theorizing, but for Helmreich they cannot yet be ethnographic subjects.

Amber Benezra, Joseph DeStefano, and Jeffrey Gordon, in "Anthropology of Microbes," envision approaching "indigenous microbial populations" by "incorporating anthropological analyses into the design and interpretation of studies of human microbial ecology" (2012). The traffic here is decidedly two-way: microbes shift "fundamental questions of relatedness, selfhood, and social transformation that have long been, and still remain, central to anthropological study," while "investigating microbes from an ethnographic perspective should provide anthropologists with new perspectives about how human biology and social practices are inextricable" (Benezra, DeStefano, and Gordon 2012). The analytical potential here is great: "We believe that negotiating the distinct and sometimes divergent methods, vocabularies, and conceptual categories that exist between anthropology and human microbial ecology is a timely and worthwhile challenge" (Benezra, DeStefano, and Gordon 2012). But this potential is limited by a decided delineation of the "social" on one side and the biological on the other; such boundary work impedes a recognition of the social character of bacterial labor.

The impediment here involves fundamental assumptions about work and labor that are integral to formulations of cultural analysis. These can be glimpsed in Hannah Arendt's classic work, *The Human Condition*. When Arendt argues for the fundamentally social character of human labor, she does so via the contrasting figure of *animal laborans*—an entity bearing "the burden of biological life," whose daily activity is entirely consumed by the struggle to maintain itself: "It is indeed the mark of all laboring that it leaves nothing behind, that the result of its effort is almost as quickly consumed as the effort is spent" (1998, 22). Such creatures exert themselves in isolation: "a being laboring in complete solitude would not be human" (22). Arendt distinguished labor from work in order to contrast the human with nonhuman; in commenting on Greek views of the nonhuman nature of slave labor, Arendt underscored, "What men share with all other forms of animal life was not considered to be human" (84)—never ending exertions, quickly consumed, leaving nothing of permanence; perpetual, in an effort simply to sustain life. This in contrast with *Homo faber*, man-the-tool-user, whose work bursts the bounds of "nature," transforming its resources and materiality into durable forms, stable spaces, and lasting buildings.

The figure of *Homo faber* represents the myth of "man-the-tool-user": a fantasy of human uniqueness that has been thoroughly undermined by a wealth of ethological examples of nonhumans devising an array of tools to facilitate routine activities. But it usefully renders a key aspect of this division between labor and work—the latter is reserved for the human because of its social character; the former is assigned to nonhumans since it lacks this quality. When Benezra, DeStefano, and Gordon consider bacteria, they begin to glimpse an end to this delineation: "We are seeing ourselves with increasing definition as a 'supra-organism' composed of microbial and human cells." In this perspective, "our microbial communities provide snapshots of those with whom we have lived, the diversity of our daily habits, as well as the impact of our changing lifestyles. . . . Gut microbial communities in humans are shared among family members and underscore the long-lasting impacts of our interpersonal relations" (2012). This vision is social in the analytical sense of breaking down the ideological unit of the individual, showing once again the verity of the anthropological axiom that a "thing" is always a nexus of relationships (Strathern 1992).

And yet Benezra, DeStefano, and Gordon too quickly undermine the potential opening presented by this view of a supraorganism by reinscribing the social on the human side of the equation when imagining what all this means for cultural analysis. As in, characterizing kinship studies as "the investigation of the social and biological associations between people that constitute relatedness," or "how microbes affect human social, political and economic life, with the primary focus being on infectious diseases," stressing that "food occupies a unique role in human lives, intersecting social with biological needs," and suggesting that microbiome scientists recognize "that there are social relations between people (beyond biological kinship) that need to be considered in designing and interpreting observational and interventional studies that target the microbiome" (Benezra, DeStefano, and Gordon 2012). All crucial points in this very important statement about engaging bacteria ethnographically. But in concert, these statements reveal boundary work whereby sociality remains exclusively on the human side of the division with nonhumans, even if our companion species can been seen, in this frame, to influence and even complicate conceptions of the "biological-social self." This delineation reflects an enduring distinction between labor and work in social theory concerning who or what can stand as an ethnographic subject, based on how we differentially regard and conceptualize human and nonhuman social activities.

Missing in both of these anthropological considerations of microbes is something that has been the source of enthusiastic theorizing among molecular biologists like Rasika and Jonathan. To pursue anthropological questions about the sociality of bacteria, I had to turn to natural science, provided in summary form by "The Social Lives of Microbes" in the *Annual Review of Ecology, Evolution, and Systematics* (S. West et al. 2007). The authors open by recounting how, over the past two decades, the view of bacteria as living "relatively independent unicellular lives" has been completely overturned. Research now reveals that "microbes indulge in a variety of social behaviors involving complex systems of cooperation, communication, and synchronization" (53). Bacteria present an important opening to rethink sociality across species lines, beyond its initial formulation "largely developed to explain known behaviors in animals such as insects, mammals and birds. The huge variety of social behaviors discovered in microbes offers a unique opportunity to test how generally that theory can be applied to other taxa" (54). The list of cooperative and communicative behaviors is quite long, but they clump into several clear clusters: quorum sensing, biofilms, and kinship.

Quorum sensing involves gene regulation in response to fluctuations in cell-population densities as groups form and expand or contract. Signaling molecules "enable single cells to assess the number of bacteria (cell density) so that the population as a whole can make a coordinated response" (S. West et al. 2007, 60–61). This chemical communication occurs within and between bacterial species, which entails "bacterial cross talk." Some species use multiple chemical signals selectively, through regulatory circuits that integrate and process sensory information; these signals are distinctive and differentiate species within larger groupings. Biofilms are structured multicellular communities and are quite ubiquitous, "being found in such diverse environments as dental plaque, wounds, rock surfaces, and at the bottom of rivers" (62). The ones on our teeth may contain as many as five hundred species. A great deal of cooperation is involved in producing the *infrastructure* of biofilms, which are materialized as an extracellular matrix, composed of proteins, polysaccharides, and nucleic acids. Typically these are multispecies communities featuring emergent functions and capacities, such as heightened tolerance against antibiotics, increased virulence in infections, and protozoan grazing. These cooperative forms are undergirded by kinship; biofilms are often initiated through kin selection, and higher degrees of relatedness favors cooperative quorum sensing and may produce higher virulence. Notably, there is no discussion of swarming in this article, as this subject was just coming into focus through the efforts of researchers like Rasika.

Efforts to move cultural anthropology "beyond the human" should find such subjects tantalizing and generative. But there is, as yet, little ethnographic engagement with this biological literature on the social lives of bacteria. Why? An initial answer is easy to reach: the cooperative behaviors reviewed by Stuart West and colleagues are entirely lodged within an evolutionary perspective and the discourse of "microbial markets." This framework produces statements that ideologically reproduce a market mentality, naturalize manufacturing, and promote an idea of rational actors, as in "the most common form of social behavior in microbes is the production of public goods. Public goods are products manufactured by an individual that can then be utilized by the individual or its neighbor" (S. West et al. 2007, 56). When evolution is seen as operating solely in terms of individual fitness, such behaviors are construed as a problem: "Public goods lead to the problem of cooperation because they are metabolically costly to the individual to produce but provide benefits to all" (56). Relatedness, in this framework, is more deterministic than anthropological notions of kinship, but it is generated as findings through the manipulation of culture: "The scale at which interactions occur will vary in nature and could be manipulated experimentally, through factors such as shaken or unshaken liquid cultures, agar plates with a variable agar concentration, or *in vivo*" (68). Rather than construing these engagements as "social constructions," "culture" here provides a point of access for cultural anthropologists to offer up new ways of theorizing or analyzing "the scale of interactions," and for thinking of evolution in less reductive terms, exactly highlighting the work of sociality.

But cultural anthropologists have as much to learn from as to contribute to research of the social lives of bacteria. This is most apparent in the recent surge of interest in multispecies ethnography. "Multispecies" has an independent career in the natural sciences, one that predates its current invocations in cultural anthropology. There is currently much debate among microbiologists over how to study multispecies biofilms, and there is an open question of how best to formulate and conduct "appropriate experiments for studying microbial interactions in these complex communities" (Røder, Sørensen, and Burmølle 2016, 503). In a recent review, Henriette L. Røder, Søren J. Sørensen, and Mette Burmølle characterize "the challenges of establishing model systems for study," while preserving or re-creating "the relevant microenvironment" (503) in the lab. In addition to their physiochemical conditions, the authors note, "bacterial communities are also shaped by the social conditions in their local environment, including *interactions within and across bacterial species boundaries* as well as interactions with higher organisms" (504; emphasis added). The artificialness of such lab settings can certainly be subject to critique (as products of labor, social and ideological), but in terms of formulating questions and answers about the work of multispecies communities that might reach beyond the confines of cultural anthropology, the opportunity here to think about scale in relation to negotiations of species boundaries seems far more intriguing.

Conclusions

Culture works across a variety of scales and so should cultural analysis; in doing so it can transgress the long-standing boundary work in social analysis of opposing nature and work. One means of pursuing cultural analysis in relation to bacteria is to focus on the crosscutting forms of attention to boundary work, a phrase that has resonance in at least three domains: in thermodynamics, in science and technology studies, and in sociology broadly. In thermodynamics, bounded systems restrict the movement of a mass while allowing energy to transfer across the boundary layer; different modes of work (electrical, shaft, or boundary) effect transfers to and from the system. In science and technology studies, boundary work names the various forms of demarcation that delineate insiders and outsiders to science (Gieryn 1983). And in sociology, this covers "a whole range of general social processes present across a wide variety of apparently unrelated phenomena—processes such as boundary-work, boundary crossing, boundaries shifting, and the territorialization, politicization, relocation, and institutionalization of boundaries" (Lamont and Molnár 2002, 168). Cultural anthropologists might bring to these projects a heightened attention to the interpretive work boundaries entail—in their perception and assertion, reproduction, and contestations. This interpretive dimension is regarded rather mechanically with swarming bacteria, as in efforts by Rasika and Jonathan to understand the transformative surface sensing capacities of bacterial flagella. But this could as well be an organic matter of the transformative power of sociality, which conditions and orients the interpretive work of its subjects, whether "natural" or human.

Notes

1. Rasika M. Harshey, personal communication, May 12, 2016.

2. Marsha Lewis, personal communication, May 19, 2016.

3. Nor in Marx's definition of commodity fetishism, by which "the mysterious character of the commodity-form consists thereof simply in the fact that the commodity reflects the social characteristics of men's own labor as objective characteristics of the products of labour themselves, as socio-natural properties of these things" (1976, 164–65).

4. Jonathan Partridge, personal communication, August 16, 2016.

5. Keep in mind that humans are not the only species that domesticates other creatures for farming (see "Domestication," in Hartigan 2014).

6. Rasika's work on swarming orients toward "emergent" aspects of "collective dynamics," so she collaborates with physicists and mathematicians, such as Gil Ariel.

7. On the resonant, multiple meanings and orientations of "rendering," see Myers 2015.

CHAPTER TEN

Rhapsody in the Forest
Wild Mushrooms and the Multispecies Multitude

SHIHO SATSUKA

Rendezvous with the Princess

"Oh, my princess! Thank you for waiting for me. I've been longing to see you for an entire year!" Mr. Kenji Oyama's face beamed and his eyes sparkled.[1] We were sitting in the dining room of his comfortable house in Banff, Alberta, a resort town in the Canadian Rockies, while he was explaining to me how he felt when he found a matsutake mushroom. It was clear that his sensation of finding a matsutake in the forest was euphoric, almost ecstatic. Oyama-san was in his early fifties at the time he told me this story in 2008. He had been running a guide company specializing in Japanese-language tours ever since he immigrated to Canada in the 1980s. He was a passionate weekend matsutake picker and well known in Banff's Japanese community as a matsutake hunting expert. On a weekend in September, he would wake up at 3:00 a.m. and drive across the Rockies six hours straight to a provincial forest in British Columbia. He said,

> After hours of driving, when approaching the forest, I see the ray of the morning light shining through the trees. That landscape coincides with my emotions. The shift from the darkness of the night to the brightness of the day reminds me that I have managed to survive one more summer business season. Running a guide company has become tougher and tougher these days because the Japanese economy has been experiencing a long recession, and we have fewer customers from Japan. But when I see the morning sun through the forest in the early fall, I see the light finally after the long hard days of struggling. Every year, I look forward to that day. When I arrive at a familiar spot and find my first matsutake of the season,

my heart beats rapidly with excitement. I feel as if I were Hikoboshi meeting Orihime.

Oyama-san described his love of matsutake by referring to the legend of "Tanabata," in which the star Hikoboshi (or Altair) is allowed to meet another star, Orihime (or Vega), only once a year across the Milky Way. The "Tanabata" legend, which originated in the ancient Chinese story of the Cow Herder and the Weaving Princess, had many variations as it spread widely across East Asia over millennia. According to a common Japanese interpretation of the story, Hikoboshi, the Cow Herder, was allowed to marry Orihime, the Weaving Princess, as a reward for his hard work and despite their class differences. Yet once they were married, the young couple indulged themselves and stopped working. Orihime's father, the God-Lord of Hosts, became angry and separated them, relegating each to one side of the Milky Way. Their deep sorrow prevented them from doing anything. They wept all day, every day. Orihime stopped weaving, so the clothes of gods wore out. Hikoboshi stopped caring for the cows, so they became thin and sick. Orihime's father felt pity and permitted them to meet once a year on July 7 if they worked hard every other day (NAOJ, n.d.; my translation).[2]

Just like Hikoboshi the Cow Herder, Oyama-san worked hard throughout the year, yearning for the day that he would see his princess. In the legend, the success of the rendezvous is contingent on the weather, and if it rains, the two lovers miss their one chance that year. Similarly, Oyama-san's chance of seeing his princess was not guaranteed: although he expected his matsutake to arrive on the day he visited the forest, whether or not he could meet his princess depended on the weather, climate, and the presence of other pickers—humans, deer, or squirrels—who might come early and take the mushroom. Oyama-san tuned his senses to the air, light, and ground and decided which part of the forest he would visit on a particular day in order to increase his chances of seeing his princesses.

However, unlike the Cow Herder, whose love was for only one princess, Oyama-san had many princesses. He showed me a photograph of matsutake he had picked within about two hours in the previous year. In the photo, the mushrooms filled the whole surface of his large dining table—the number was easily over one hundred. Despite this large number, he insisted that he felt special affection for each mushroom. He told me that he remembered vividly how he met each "princess." Every single encounter with a mushroom was unique and precious to him. The distinct feeling of the stem when he first touched a mushroom still stayed with him. Same with the fresh aroma that evaporated from its body. He remembered the moisture in the air, the sound of the wind, and the chatter of the squirrels and birds that surrounded each rendezvous. By closing his eyes, he summoned those feelings back. He could visualize the angle of the light, the contour of the forest floor, and the color of the mossy bump in which the matsutake hid. He would gently peel the moss to reveal the matsutake's round white cap. His senses were keenly attuned to the fungus and its habitat.

What do the stories of matsutake enthusiasts like Oyama-san tell us about "how nature works"? His story invites us to notice a relationship between the human and nonhuman that is different from the hegemonic approach to nature in capitalist society. Nature is not merely a passive object that humans work upon or a resource from which to extract materials useful for humans. Rather, his story suggests that people are still making intricate connections with nature by closely attuning their senses to other forms of life. The charismatic characteristics of the matsutake mushroom help remind people of these connections and that humans have cultivated particular senses to tune in to the rhythms of other beings on the earth. Here, nature is not external to humans. Humans are part of what is now called "nature," or the multispecies multitude.[3] This chapter explores how the elusive charisma of the matsutake mushroom might open up people's senses toward how nature works.

Matsutake as a Guide

Matsutake are a good guide for responding to the call for this volume: to trouble conventional perceptions of "nature" and "work" (Besky and Blanchette, this volume; see also Haraway 2016). The matsutake is a mushroom long treasured in Japan as an autumn delicacy. It is thought to represent the ethos of the Japanese culinary tradition, which is the appreciation of nature by eating seasonal items. Its significance is often described as similar to the truffle in French cuisine.

In English literature, the matsutake is described as a "wild" mushroom (e.g., Hosford et al. 1997; Pilz and Molina 1994) because humans have not yet found a way to cultivate it artificially. For over a century, many attempts have been made, yet none have succeeded. Like the truffle, chanterelle, porcini, and most of the highly valued mushrooms in the world, matsutake are mycorrhizal.

Unlike saprobic fungi—such as common button mushrooms and shiitake—that grow on decayed organic matter, mycorrhizal mushrooms require a symbiotic relationship with living host trees. While scientists know that matsutake need to exchange nutrients with their host, the exact mechanism puzzles them (Suzuki 2005). It is hard to reconstruct the complex interactions in the symbiosis as there are many factors to coordinate. It is not only about creating the ideal conditions for one organism but a particular coordination among a variety of organisms so that they form relationships necessary for the fungus to produce a fruiting body. The high price of mycorrhizal mushrooms in the market reflects this difficulty. As humans do not know how to cultivate the matsutake, the mushrooms need to be foraged in the "wild" forest.

Yet one of the pioneering matsutake scientists, Makoto Ogawa, describes matsutake as mushrooms that "human beings have unintentionally cultivated" (1978, 20; my translation). Matsutake scientists generally agree that the matsutake is a weak competitor among other fungi and microbes in the soil; if the soil is rich enough to provide food for other species, matsutake cannot thrive. Thriving happens in specific coordinated ecologies. Matsutake's main host in central Japan is the red pine. Ecologists categorize the red pine as a "pioneer species" among trees, which thrive in nutrient-poor soil generated after ecological "disturbance." The disturbance can take place naturally, such as forest fires or volcano eruptions, or by human intervention, such as clearing the existing forest. In nutrient-poor soil, the red pine and matsutake form a reciprocal relationship: the matsutake offers minerals in a form that the pine can intake, and, in return, it receives carbohydrates produced by the pine through photosynthesis. Matsutake's prime habitat in mainland Japan has been in *satoyama*, secondary forests near agricultural settlements. Historically, satoyama played an important role in providing firewood and green manure. People regularly went to the forest, selectively cut trees, and cleared the forest ground. By doing so, they unintentionally maintained the satoyama forests and created a niche for matsutake to grow in.

Matsutake habitat is a product of a long history of entangled cohabitation among the fungus, pine, humans, and other beings. However, since the 1950s, satoyama landscapes have been transformed. Many critics argue that the main cause of the transformation was the "fuel revolution" (e.g., Arioka 1997; Saito and Mitsumata 2008). With the introduction of imported fossil fuels, people stopped collecting firewood in the forest. At the same time, Japan experienced rapid industrialization. The population drained from rural agricultural communities to urban cities and semi-urban industrial areas. Accordingly, an increasing number of humans have become disconnected from satoyama, leaving them unattended. Thus, in Japan, people stopped managing the satoyama, the practice of unintended cultivation of matsutake mushrooms, and the mushroom lost its habitat.

Meanwhile, since the 1960s, as the matsutake lost its habitat and the harvest drastically declined, its price went up sharply. Since the 1980s, in order to substitute for an expensive domestic matsutake, Japanese traders began importing them from overseas. The matsutake is arguably one of the most expensive "wild" mushrooms (Alexander et al. 2002; Mortimer et al. 2012). As such, it has attracted many harvesters and traders across the globe. Now, over 95 percent of matsutake consumed in Japan are imported from many countries, including China, Korea, Canada, the United States, Mexico, Bhutan, Turkey, Morocco, Sweden, and more.[4] In those countries, except Korea, matsutake were not necessarily favored until the Japanese started to buy them at a high price. The value of the matsutake was realized through global trade.

In Japan, the loss of the matsutake's habitat symbolizes the severe damage caused by industrialization to both environment and society. It represents serious problems concerning agriculture and forestry and the heightened social concerns with Japan's heavy dependence on food and timber imports. Due to their scarcity and high price, most of the remaining matsutake-producing forests in Japan have been tightly controlled by local forestry communities. During the matsutake harvesting season, some communities hire guards to patrol the area and keep outsiders out. In contrast, in many other countries where matsutake have recently drawn people's attention, the "discovery" of exporting a non-timber forest product brought excitement in various ways. In Yunnan, China, "matsutake mansions" have been built using the profits from the mushroom trade (Arora 2008; Yeh 2000). In Bhutan, a national park created a matsutake festival to attract Japanese tourists. In Oregon, people gathered for the "white gold rush" to collect matsutake (Guin 1997; Tsing 2015). In northern British Columbia, the matsutake is considered an item for sustainable development on First Nations' lands (Collier and Hobby 2010; Menzies 2006). In this context, in the United States and Canada, recreational matsutake picking has been expanding among Japanese immigrants. In fact, the global expansion of matsutake trade reveals that matsutake and its closely related species are quite common across the northern hemisphere, as the fungi found habitats in a variety of human-disturbed landscapes. Matsutake have also stimulated scientific interests in understanding the puzzles of symbiosis. These entangled relations among matsutake, trees, and humans prompt us to question the distinction between "wild" and "domesticated," as well as "nature" and "culture."

Multispecies Multitude

"Nature" needs attention here. The word "nature" is conventionally used as a singular noun in contemporary English. It generally means the material environment and its components, such as plants, animals, microbes, and minerals. It is an abstract concept that covers all that is not artificially made by humans. As Raymond Williams points out, this conventional usage of "nature" as an abstract singular noun emerged in the Enlightenment as a generalization made by modern scientific rationality, making the concept a product of a specific history. As he states, "The extraordinary accumulation of knowledge about actual evolutionary processes, and about the highly variable relations between organisms and their environments including other organisms, was again, astonishingly, generalized to a singular name" (1976, 224). This abstract generalization of nature leads to an imaginary of nature as a collective of things external to human beings and serves as the foundation of scientific objectification, as well as the perception of nature as a resource for capitalist development. But "nature" is an enigmatic concept. How are these different beings placed under this single category and given common membership? This notion of nature also entails further tensions as it is unclear whether nature is external or internal to the human, and its boundary is often ambiguous. While being ambiguous, it claims universality; "nature" exists everywhere and is shared by all people even though this abstract generalization developed in specific epistemological and ontological traditions.

Anthropologists have documented numerous ethnographic examples that challenge the universality of this idea of nature (see, for example, Descola and Palsson 1996; Strathern 1980; Viveiros de Castro 1998). As I have discussed elsewhere (Satsuka 2015), the way in which the English word "nature" was introduced in Japan in the nineteenth century elucidates that "nature" was one of the most challenging concepts to translate as there was no equivalent in existing East Asian epistemology. Yet the translation of "nature" was central to the Western-modeled "modernization" process as a response to the Euro-American colonial expansion in the region (see also Liu 1995). After decades of debate, Japanese intellectuals settled on the translation of "nature" to "*shizen*," yet the conflict between "nature" and "shizen" was not solved. Although both words connote something that is in contrast to human intentions, "nature," as a singular noun indicating the whole material world external to and in opposition to the human, is not equal to "shizen," derived from the Taoist conception of a relational, natural state of being, internal to and integrated in the human realm. The tension is contained in the word and has continually generated indeterminacy. It also continues to shape the lives of people in the twenty-first century as the universal claim of "nature" attracts people while its indeterminacy shapes politics among different understandings of, and practices with, nature. Oyama-san's employees, the Japanese-speaking nature guides, have been working at the frontline of this negotiation as they translate the Canadian National Parks' idea of nature to Japanese tourists. Their everyday practices have been shaped by this tension and indeterminacy in the translation of nature.

"Work" also requires closer consideration. Obviously, the matsutake "works" upon Oyama-san. It is the fungus that attracts the human. It affectively and physically moves him all the way across the Rocky Mountains. We could describe the way he is attracted by the matsutake as an example of "how nature works" upon humans. It suggests that work as subjective action is not the privilege of humans. But what is happening here is not a mere reversal of subject and object, or actor and the one acted upon, between humans and nonhumans. It is not enough to say that the fungus is an actor that works upon humans. Rather, what we need to pay attention to is the communicative practices of forming a relationship that co-constitutes humans and fungi in an entangled web of life. Further, "work" here is not only concerned with functional relationships. It is possible to interpret that the fungi attract humans (and other animals) because they help the fungi reproduce as they spread the fungi's spores on the forest ground. But instead of reducing the fungus-human relationship to the efficiency of material circulation and utilitarian function, it might be helpful to consider it an act of forming a relationship, closer to "action" in Hannah Arendt's distinction between labor, work, and action (1998). Since humans cannot live without the existence of other beings, we might want to extend "human conditions."

In order to understand work as communicative practice, retooling the notion of the "multitude" would be useful; specifically, by including nonhuman actors and extending its application to multispecies relationships. Multitude is a concept drawn from Baruch Spinoza and reintroduced for theorizing the social and political relations of late capitalism—often labeled "post-Fordist" capitalism—by Italian Marxists (e.g., Hardt and Negri 2004; Negri 1989). Among the discussions of the multitude, I found one by Paolo Virno (2004) most useful, as he focuses on the communication process at the heart of "work" and explains the inseparability of labor, work, and action. The multitude indicates a collectivity that does not presuppose unity, as does the older category of "people," or national or class identity. The multitude denotes a plurality, a mode of being that remains "the many" without being converted into a synthesized unity. Virno explains that the multitude caused a sense of uneasiness for seventeenth-century political thinkers, such as Thomas Hobbes, because the multitude was considered anti-state and against the "people" who transfer their natural rights to the sovereignty of the state and allow the state to act as their representation (2004, 23). The concrete example of the multitude at that time was people in "exodus," or stateless people. In the contemporary capitalist world, Virno sees the multitude in the workers of the globalizing economy. Contemporary workers become subjects of the movement of capital that extends beyond national boundaries. Meanwhile, the expansion of value production in late capitalism foregrounds the centrality of intangible aspects of labor, as contemporary capitalism makes explicit that value production relies heavily on information and communication, which is produced by intellectual, cultural, and affective labor. In Virno's rereading of Arendt, he understands the work of one's general intellect as the human condition. Instead of seeing public political action and labor as separate, he sees political space possible at the heart of capitalist production. Because general intellect is inseparable from labor, the site of labor can be a "common place," where people communicate and negotiate. Labor requires the use of general intellect in the presence of the other, which is the basis of political action. Yet, unlike more conventional Marxist theorists, Virno does not assume a teleological and unidirectional evolutionary progress. Because of its multiplicity, the multitude resists the foreclosure of identity and politics, whether emancipatory or not. The subjects who form the multitude consist of a permanent interweaving of relationships and individuated characteristics, or rather, "the subject is this interweaving" (Virno 2004, 78). The interweaving is the result of productive cooperation by the general intellect, yet it does not necessarily mean it is a peaceful process. Rather, it engenders crisis and conflict. Drawing from Gilbert Simondon, Virno describes the multitude as an "amphibian" subject, both inside and outside of the capitalist system, simultaneously obedient and resistant.

Virno's multitude urges us to pay close attention to the constant process of communication and the contingent formation of relationships among various actors from which subjectivity and politics emerge. Yet his discussion is limited to human relationships. What if we extend this process to human-nonhuman relationships? What if we free our understanding of communication from verbal and intentional communication among humans? It seems the matsutake enthusiasts like Oyama-san have already been developing a method of communication with the fungus with their senses. They seem to form the multispecies multitude contingently and unintentionally. Obviously, the communication among the multispecies multitude does not rely on conventional human language. Central to this communication process is the attunement to the rhythms of life. Because of their elusiveness, wild mushrooms like matsutake are good guides for humans to hone their senses to the rhythms of other beings.

Attunement as Multispecies Communication

People who went matsutake hunting with Oyama-san told me that he had a special "antenna," or exceptional instinct, built into his body. They said that he could detect matsutake while driving a car on a dirt road at one hundred kilometers per hour: he would suddenly stop the car, drive in reverse about one hundred meters, pull over, hurry into the forest, and return with some mushrooms. The story sounded unreal to me because matsutake are not easy mushrooms to find, as they are often covered by moss in western Canada. But I had the opportunity to witness Oyama-san's use of his "antenna" when he took me matsutake picking with his employees in the fall of 2016. Just before dawn, while rushing to catch the first ferry across the river, he suddenly slowed the car, found a roadside bush, pulled over, walked into the bush, and soon came back with a large mushroom. How could he find a mushroom in the dark from a moving car in the monotonous-looking roadside rows of conifers that he could visit only several times a year? How could he coordinate his senses with the mushroom? It looked like magic. A sense of wonder followed the mushroom into the car as we were fascinated by the masterful tuning of his senses to the forest.

Of course, Oyama-san remembered where he found matsutake in previous years. Still, he had to detect which particular patch to go to among the numerous sites he knew. He tuned his senses to the temperature, moisture, light, and feeling of the air to imagine where in the forests the fungus produced the mushroom. He told me that this "primitive" (*genshitekina*) form of communication with the mushroom is the attraction of matsutake hunting.

It was not only Oyama-san who was caught by matsutake. The "matsutake fever" seems to be contagious and developing over the past few years. When I first lived in Banff in 2000, conducting fieldwork on Japanese nature tourism, most of Oyama-san's employees, who were in their twenties and early thirties, were not so enthusiastic about matsutake picking. Oyama-san's strange hobby was the target of mockery among them. They often teased him about how he was mesmerized by the wild mushroom. Yet, over a decade, I have observed that many of those who worked with him became serious pickers themselves. How do we make sense of people's fascination with matsutake—or rather, the fascinating power of matsutake to attract people?

Part of the matsutake's attraction is its scarcity in Japan. Oyama-san and most of his employees did not have much experience picking matsutake in Japan—some had never seen the mushroom in the forest. They learned how to pick matsutake after moving to Canada. Most of the popular recreational picking areas in southwestern Canada are on public land, where people can enter the forest freely. But the matsutake's attraction seems to be more than its scarcity in Japan contrasted with the bountifulness and freedom of harvesting in Canada. One of the hiking guides sent me an email, attaching a photo of matsutake she found while she was wandering in the forest between tours. She told me that once she started to collect matsutake, the forest looked different. As an outdoor enthusiast, she has been walking in the forest every day for over fifteen years. For work, she takes tourists along hiking trails, showing them the spectacular photogenic mountains and lakes in protected areas in national and provincial parks. But once she learned to tune her senses to find matsutake, she started to experience a sense of wonder in the mundane landscape.[5] When she found matsutake, even in the "ruin" of a clear-cut area, the feeling of joy prevailed over the sadness of seeing the destroyed forest. She sensed the existence of different worlds lived by a variety of visible and invisible beings above and beyond the apparent crude relationship between trees and humans in industrial timber production and transportation. Even though these worlds evade the human sensibility disciplined in contemporary industrial society, they surely exist.

The stories of matsutake enthusiasts suggest that central to matsutake's charisma is its elusiveness and unpredictable temporality. With its elusiveness, the matsutake leads humans to a different world, by opening a window to a different spatio-temporal domain lived by other beings. The elusiveness requires "attunement" (Choy and Zee 2015), or the art of "noticing" (Tsai et al. 2016; Tsing 2015, 2019). It guides pickers to attune their senses keenly to the life world of matsutake and makes them realize how the fungi coordinate their tempos and rhythms with those of other beings that compose the forest landscape. From this point of view, a matsutake enthusiast's tales, such as Oyama-san's, can be understood as an interspecies encounter. It can be told as a story of the prowess of matsutake to attract humans, rather than attributing the communication skills exclusively to humans. It is a story of the matsutake's ability to communicate with other species and draw them into the rhythm of the matsutake's life. This communication does not rely on language and intentionality (Satsuka 2019). Attunement to this rhythm is central in interspecies communication. By being drawn to the rhythm of matsutake and bringing the mushrooms around, humans (and other animals, such as squirrels and deer) have been unintentionally collaborating with the fungi and helping the mushrooms' worlding practices.

Rhythms of Rhapsody

German philosopher and psychologist Ludwig Klages argues that rhythm is the most general phenomenon of life experienced by all living creatures, including humans (2011, 14). Klages's notion of rhythm is useful in exploring multispecies relations when combined with contemporary Jakob von Uexküll's notion of *umwelt* (2010), the world perceived by a specific organism based on its particular sensory structure and functions. Each organism perceives its environment with its specific sensory system, and in turn, acts on the environment with its particular mode of action. Therefore, the way organisms experience the world is unique. The life world of each organism is lived differently with rhythms particular to that life form.

Klages explains that the word "rhythm" comes from Greek "*rheum,*" literally "that which flows." For him, rhythm is a ceaseless flow that cannot be divided into intervals. He points out that tempo and rhythm are often confused, but the regularity of tempo stands in contrast to rhythm, as the former is consciously produced by human intellect and will, while the latter is an unconscious expression of vital life itself. Rhythm and tempo are not mutually exclusive, and they can fuse with one another. Rhythm can be intensified by tempo. However, while rhythm can appear in its complete form in entire absence of regularity and tempo, tempo cannot appear without the cooperative mediation of rhythm. The decisive difference is that "rhythm represents a 'renewal of the similar' while the regularity or tempo represents 'a repetition of the same'" (Blasius 1976, 65). Tempo can be used to control the vital rhythm of organisms and plays an integral role in regimenting time.

Regimentation of time is one of the key features of labor in capitalist production (see Thompson 1967; Lefebvre [1992] 2013). Workers in factories, stores, and offices are disciplined to manage their time and to coordinate their sensory and bodily practices with certain speeds, tempos, and cycles with other workers to produce necessary effects and affects for making profits in time. Many thinkers have critiqued how regimenting temporality has alienated workers from their own lives in order to coordinate the tempo of their labor in the capitalist production system. These critiques are useful in understanding capitalist attempts to mechanize the rhythms of life, but their arguments seem to be heavily shaped by the concerns of industrial capitalism led by large corporations developed from the late nineteenth century to the twentieth century. The irony of capitalist industrial production is that it requires "natural" resources—including human resources—that the capitalist system itself cannot reproduce. It relies on nature's production and on the existence of resources external to the system. What needs to be built onto these previous discussions is the realization that those "natural" resources are the products of the flow of life, which cannot be completely controlled by the mechanized tempo of capitalist production systems. The temporal coordination of work poses extra challenges in industries in which the work of nonhuman species is integral, such as agriculture and forestry, because the temporality of other species does not necessarily match with that of humans.

Anna Tsing (2015) suggests that one of the challenges of telling such a complex tangle of temporalities is how to tune our senses to histories at different scales. Multispecies history involves an evolutionary scale of millions of years, human environmental change with a scale of 1,500 years, a mushroom's development scale of less than a hundred, and a human social change scale of the last thirty years. We need to hear different melodic lines played by different species. For this purpose, she uses the "fugue," a baroque music genre, as a metaphor. While Tsing highlights multiple temporal scales in natural history, my analysis focuses on multiple rhythms that compose multispecies communication, the moments of encounters when one species is drawn into the rhythm of another.

In order to examine the contingent resonances and dissonances of rhythms emanating from multiple beings, I found rhapsody is a useful working metaphor. In rhapsody, we hear certain melodic lines with particular rhythms abruptly followed by something else, yet the different lines often come back after a while. Even though music as sonic representation works with the human sensory system that can capture a certain amount and length of vibrating waves in the air, rhapsody creates in listeners the trace of rhythms that are not present at a particular moment. It has the potential to direct our attention to the spatio-temporal dimension simultaneously existing with what is physically presented to the human sensory functions and to the possibility of communication beyond the modern metaphysics of representation (see Derrida 1982). Rhapsody's eclectic, fragmented cacophony brings our attention to the multiple beings existing in the world with various rhythms. It helps us to attune to multiple rhythms simultaneously happening although not being realized. A fruiting body of a mushroom is a product of a rhapsodic jumble of multiple rhythms, coincidentally resonating with the work of trees, fungi, insects, and humans as well as climate change and the global political economy. Rhapsody expresses a rhythm of the multispecies multitude.

Entangled Temporalities and Values

When I first heard of Oyama-san's interest in matsutake, I suspected that he was looking for a business opportunity. Because Japanese souvenir stores and companies sell dried Canadian matsutake to Japanese tourists, and courier companies have started gift services to send Canadian matsutake to Japan, I thought Oyama-san also wanted to jump into the matsutake business using, but also substituting, his declining tourism business. But it turned out that he had no interest in matsutake as a business. Rather, matsutake hunting was solely his recreation. For him, the value of the matsutake is that this mushroom helped him forget about the business world. He found pleasure in moving into an "other world" (*betsu no sekai*) and being able to "forget time" (*jikan o wasureru*) by experiencing rhythms and temporalities different from his everyday life.

It is significant that Oyama-san used the otherworldly "Tanabata" legend to explain his matsutake-hunting experience. Arguably, the earliest remaining record of the story is found in the "Nineteen Old Poems" compiled during the Han dynasty in China (206 BC–220 CE) (Kawata 2016; Wu 2003). The story transformed into numerous versions as it incorporated local rites, rituals, belief systems, and folk customs across Asia (Kawata 2016, 178). Yet, as Ko Kawata points out, a common feature remains: its unique temporality. The annual meeting of the Weaving Princess and the Cow Herder that occurs as if they do not age or die produces the effect of eternity. The euphoric excitement of the rendezvous of the young couple is always fresh. The core attraction of this legend is that it stimulates the wish for an eternal continuation of this beautiful yet ephemeral moment (Kawata 2016, 170). It pauses time's flow through the joy of the rendezvous. The moment exists only here and now.

The sense of eternal continuity of "now" is related to another feature of the "Tanabata" legend: the connection with the other world. In the ancient versions, it is a love story between supernatural figures. Both the Weaving Princess and the Cow Herder were celestial divinities, dwellers of the sky (*ten*) represented by the stars Vega and Altair. However, since the time of the Tan dynasty in China (618–907 CE), it was synthesized with another tale that was widely spread called "Hagoromo Densetsu," or "The Robe of Heavenly Feathers" (Kawata 2016, 165), and the story transformed into one bridging the heavenly sky and the mundane earthly world. The common plot in this later version is that beautiful young women came down to earth from the sky to bathe in a spring. A humble man (whose occupation varies from a cow herder to a peasant to a lumberjack) hid one of the women's robes. Without the robe, the woman was not able to return to the sky, and so she married the man. After giving birth to his children, the woman found the robe and went back to the sky. The man was allowed to meet the woman up in the sky, but only once a year; a motherly figure helped the couple rendezvous. Kawata argues that the transformation of the "Tanabata" legend from this period onward reflects people's ambivalent feelings toward the development of patriarchal civilization; the variations oscillate between male domination over women represented by the Weaving Princess's father and the Cow Herder's act of hiding the Weaving Princess's robe to marry her and dependence on women for reproduction and affective relations represented by the caring characteristics of the Weaving Princess and the helpful motherly figure (153).

The "Tanabata" legend indicates the entanglement of different temporalities: the processual flow of time in this world and the eternal continuity of "now" in otherworldly space. These are associated with different values. The value of productivity is generated by the Cow Herder's and the Weaving Princess's everyday engagement with their work as processes of producing food or clothes. As their work is accumulated through time, the end products will be made and valued, whereas the affective value of euphoric excitement derives from the encounter with otherworldly beauty. The rendezvous stops the flow of time, continually re-creating the person in the "now" of the encounter.

The entangled temporality of the "Tanabata" legend suggests an interesting tension in temporalities we experience. As Laura Bear points out, "Institutions mediate divergent representations, techniques, and rhythms of human and nonhuman time" (2014, 6); yet there is always instability between the institutional effort of managing and the actual practices of time experienced in people's everyday lives. In that instability, as Matt Hodges suggests, we can find other temporalities as "sideshadows" (2014, 39, 48n14) along with hegemonic "processual" time in capitalist production (36).

Oyama-san's recreation time was mediated by the institutionalized temporal management that separates his work and leisure time. During the matsutake season, he normally had one day off a week if he was lucky. His leisure time was limited, and he had to calculate the hours needed for driving to get to the matsutake forest and come back home so that he could work as scheduled the next day. (For example, he planned to get up at 3:00 a.m., leave home at 4:00 a.m., drive six hours to the forest, start hunting at 10:00 a.m. for one hour in this patch, another hour in that patch . . . and leave around 2:00 p.m. to come home by 8:00 p.m., and so on.) However, he was also experiencing different time when he was meeting his princesses, uplifted in the "now" of having a rendezvous with beauty coming from an otherworld.

The tension between processual industrial time and the constant re-creation of "now" is experienced acutely by the nature guides hired by Oyama-san. His company prides itself on the quality of its guides' work, catering to a variety of tours, including mass-packaged commercial tours subcontracted to major Japanese travel companies and more personalized small-group hiking tours. The company trains the guides to provide "sincere" service to tourists coming to the Canadian Rockies to enjoy magnificent nature in their short vacation time. Most of the guide trainees left Japan to live in the cosmopolitan mountain town of Banff, and many of them are outdoor enthusiasts who have obtained, or seek to obtain, professional accreditation for guiding, climbing, or ski instructing. In training sessions I observed, the senior guide constantly reminded the trainees of the sense of amazement when they first encounter glacier-hung mountains and milky emerald-colored lakes (see Satsuka 2015). The trainees, as avid nature lovers, were enthusiastic in sharing with tourists moments of being mesmerized by the beauty of the natural landscape. Yet once they started to lead a tour, their "sincere" hope of sharing the "now," the uplifted moment of amazement, often faced a temporal challenge as they had to cut off the moment in order to conduct the tour efficiently. They need to maintain regimented processual time in order to take the tourists to all the spots documented in their itinerary. The documented itinerary is considered a business contract for which they are paid to cover all the items in the contract within the stated hours. The guides often expressed frustration to me that they wanted to let the tourists stay in places they seemed to be amazed by and let them enjoy it by "forgetting time," yet they had to urge the tourists to leave in order to bring them back to the hotel in time.

In the guiding business, value is produced through the complex dynamics of these contrasting temporalities—the affective value of experiencing the moment of amazement and the commercial value of offering a quality experience in a measured processual time. This value production depends on the guides' skills of balancing these different temporalities, managing the tension and communicating smoothly with the tourists. The guides are required to tune their senses to the natural landscape—flora and fauna, weather and climate—as well as to the tourists' physical, sensual, and emotional conditions and help the tourists experience an affective encounter with the natural landscape.

In his days off, Oyama-san occasionally took some of his employees matsutake hunting. At first, his employees, who mostly grew up in urban areas in Japan, had a hard time finding mushrooms. Oyama-san instructed them to sharpen their senses and tune in to the atmosphere of the forest. He often asked them, "If you were a matsutake mushroom, where in this forest would you want to be?" Look closely at the ground. Touch the soil and detect the moisture. Feel the air. Smell the forest. Listen to the wind, birds, and animals. He suggested that if a person closely tuned their senses to the environment and felt the lives of other creatures, then they would be able to find matsutake. Oyama-san found that this act of tuning their senses to other beings was helpful for young urban adults to overcome their self-centeredness, as they learned to see the world from the point of view of other forms of existence. He considered this attitude important for their personal growth and maturity. Yet this moral value was also entangled with the value produced in his business of nature guiding. Being attentive to others and the environment and developing sympathetic communication skills was important for his employees as service industry workers, especially as tour guides in the Canadian Rockies.

Guiding work is a product of communication among the guides and the human and nonhuman beings they encounter. It exemplifies the value production in the multispecies multitude. In this value production, the temporal coordination is integral, and guiding requires intricate balance between the disciplinary regimentation of processual time and the appreciation of the temporal hold—here and now—of the encounter with other beings whose lives are shaped by different rhythms. As such, the workers' lives are shaped by rhapsodic coordination of different rhythms and temporalities. The workers' communicative mediation is a site of struggle between hegemonic processual time and the other temporalities sideshadowed by regimented time. As intermediaries of these multiple temporalities, the workers who constitute the multispecies multitude dwell both inside and outside capitalism, an ambivalent space whose work is inseparable from the action that moves and transforms human and nonhuman others. As such, their work shapes the conditions and possibilities for multispecies cohabitation as well as the awkward moments of dissonance and disruption of the condition of cohabitation.

Conclusions

By following the matsutake's lead, and by attending to the experiences of the matsutake enthusiasts, what do we learn about how nature works?

First, what is labeled "nature" can be conceptualized as a multispecies multitude. It denotes that various beings on the earth exist as "the many"; they coexist but without a presupposed and foreclosed collective unitary identity. The objectified identity of "nature" given to the multitude of beings leads to the perception of these beings as a collection of "natural resources" that is supposed to exist only to be controlled and utilized by humans. Instead, "multispecies multitude" foregrounds the singularities of each being. Each existence—including human—lives with its own rhythms. These rhythms are constantly in an encounter with other rhythms, creating resonances and dissonances. The relationship among those who constitute the multispecies multitude is contingent on the ways these various rhythms are coordinated or discoordinated. The rhythms of the fungus, red pine, and human were coordinated to create a niche for matsutake in satoyama until the first half of the twentieth century in Japan. However, the coordination has been disarticulated since imported fossil fuels replaced firewood and charcoal in the process of intensified industrialization. The consequence was the spread of unattended agricultural forests as green wastelands.

Yet the matsutake with its charisma still draws people into the rhythms of the fungus. This leads to the second point about the "work" of nature, or the multispecies multitude. The work depends on interspecies communication. This communication is not necessarily intentional or verbal. Central to interspecies communication is the attunement to the rhythms of other beings. The stories of matsutake enthusiasts tell us that human beings have cultivated the skills to communicate with other beings not by human language but by tuning their senses. Their practices of attunement guide them to be aware of the worlds generated and lived by other beings, even though these worlds often evade human eyes. It also elucidates that while industrial capitalism tries to regiment time by regulating tempo, there are other temporalities operating even if they are "sideshadows" and hard to capture by human senses disciplined in this industrial regimentation. Those sideshadowed temporalities never cease to exist. In the ecstatic encounter with other beings with different rhythms, the matsutake enthusiasts experience the euphoric continuation of "now," a pause in processual time.

The matsutake, particularly with its evasiveness, suggests a tension that humans have in their multispecies relations. On the one hand, humans have been interested in controlling the rhythm of other species. They made efforts through industrial interventions and scientific experiments. On the other hand, humans have also been fascinated by the contingency and unexpected rhythms of other species that they cannot fully control. Stories of matsutake enthusiasts indicate that what is integral to multispecies cohabitation is learning how to live with these contingencies, the continuous resonances and dissonances, coordinating the flow of time, yet sometimes experiencing the astonishing moment of time paused, created by multiple rhythms among different species. The matsutake, with its rhapsodic affects, reminds people of these multispecies relations in the ruins of industrialization and of the ambivalent process of value production by the multispecies multitude.

Notes

1.   All quotes from "Kenji Oyama," personal communication, August 21–24, 2008. This is a pseudonym. Going forward, I have added "-san," a common Japanese honorific, after his last name following the customary practice of ethnographic depiction of a Japanese person to indicate the ethnographer's respect to the informant.

2.   NAOJ (n.d.) clarifies that although Orihime and Hikoboshi can meet once a

year in legend, the stars Vega and Altair are not able to meet as the distance between the two stars is 14.4 light-years.

3. I do not mean to assume that "species" is always the relevant unit to analyze the human-nonhuman relationship. I also do not intend to privilege living organisms over other beings in the relationship, as the boundary between organic and inorganic beings is not always clear, especially in thinking about relations with soil and rock. Due to limited space, I use "multispecies" as a starter and a placeholder for thinking about the complex relationships among multiple beings that constitute the world.

4. What is labeled as "matsutake" in the commercial market consists of several close but different species. Matsutake harvested in Japan, Korea, China, Bhutan, and in northern Europe (Sweden and Finland) are *Tricholoma matsutake*. Those from the Mediterranean and northern Africa are *Tricholoma caligatum*, whereas North America has mostly *Tricholoma magnivelare* (Forestry and Forest Products Research Institute, Mushroom Research Laboratory 2008).

5. See Ehlers, Fredrickson, and Berch (2007) for the particular ecology of the matsutake-growing areas in southern British Columbia.

CHAPTER ELEVEN

Kamadhenu's Last Stand

On Animal Refusal to Work

NAISARGI N. DAVE

In the spirit of refusal, I'll start with a somewhat meandering route. I've been thinking a lot about dreams recently. It began in January 2016, when I read an interview in the *New York Times* with an artist who said that one of the ten books he would take with him to a desert island would be a bound copy of a four-line nursery rhyme: "Row, row, row your boat / Gently down the stream / Merrily, merrily, merrily, merrily / Life is but a dream."

Like most of us, I had heard this rhyme dozens of times before. But suddenly it appeared to me as the very meaning of life. Of course: life is a dream, at some point we all wake up (the dream ends), and so ultimately nothing really matters all that much (except, perhaps, dreaminess). And so I began to wonder, everywhere I went—why do we waste our dream doing things that are boring, sensible, and rote?

Or perhaps the question is, why do we waste our time *interpreting* our dream in ways that make it *appear* boring, sensible, and rote?

For example, let me share a dream from my fieldwork, though it is a dream I did not really have.

It's February 2013, and I'm standing in the sunshine in the parking lot of a nondescript office park when suddenly a truck rolls in. The door of the nondescript building opened like a maw and out came several men in green jumpsuits with one little boy in regular clothes but no shoes. The truck was filled with squawking birds and the men surrounded it, haphazardly, reaching in and pulling some of the birds out, and then, with sudden alacrity, began to pull *all* the birds out and throw them into milk crates, which were carried inside so the birds could be hung upside down by their feet and sent into a stunning machine from which they came out like stiff little meaty stones with ropy necks whose veins were cut by a woman with alarming regularity and with no regard to the blood splattering all over her as the birds morphed from stiff little meaty stones to flapping, squawking art objects with bright-red blood spurting from their new orifices as they entered a vat of boiling water, after which they were still thrashing and the feathers were stripped from their hot little bodies by a man in yet another green jumpsuit, and then they came around to a row of three new men, one who announced himself as the vent cutter, which means the anus cutter, and he knifed through each anus-vent, sending brown and yellow liquid squirting everywhere including onto his face and his jacket, and then the gullet remover reached into the vent once known as an anus and pulled out the innards, and then there was a cleaner who hosed the birds down what with all the shit and innards, and then the birds went higher on the conveyer (they now all seemed dead, and I remember thinking that was good) where a man sat high above on a little platform with his legs crossed, his head nearly touching the ceiling, and he without expression except for the small glint in his eye as he made eye contact with me, pulled the feet off each bird. I look down: on the floor all around me are a thousand severed feet. They fall like rain.

I say this is a dream that I did not actually have because, indeed, it all happened. Though according to row-row-row-your-boat philosophy, of which I am an adherent, there is no meaningful difference between the dream and the not-dream: both are at times highly visceral, at others, barely registered; both come to an end; both have effects that ultimately fade; both are full of the remarkable, the surreal, the inexplicable; both express the truth of no single context but rather constitute their own. The difference is only in what we are expected to make of each, as anthropologists anyway. And this is what I have been tempted, or expected, to do with my dream/not-dream: *to put it into context*. That context begins, for me, shall we say, with an idea called "India," but which, as "ethnographic context," is taken to be as real as real can be. And that is a context in which there are far more important things than chicken feet, far more pressing matters then feverish dreams. It is a context in which chicken equals protein, and the green jumpsuited men are laborers first—laborers with creeds and castes, we mustn't forget (genders or erotic sensibilities, who cares?)—a context in which Hindu fundamentalism, which justifies communal violence against those who eat animals, makes contemplation of the lives and deaths of chickens obscene at best, enabling of atrocity at worst. In other words, it is a context in which none of this can matter. And yet it did, if you ask me, or the chicken, or the man with a glint in his eye (and it didn't). I suppose I could insist, by using context strategically ("this is about labor, about the nation-state, about postcoloniality"), that it does matter. But you know what? I don't think I want to.

My friend and colleague Amira Mittermaier (2011) was rather more game. She wrote a book about dreaming in Egypt, but in 2011, as now, the political context of Egypt was a decidedly undreamy one. And so how to write about the seemingly trivial (dreaming) in a time and place of violent consequence? Well, she would show and argue, as in the title of her book itself, that all dreams are not trivial at all; hers would be a book about *Dreams That Matter*, dreams that give us insight into things of consequence.

I have no problem with insight, but I am increasingly skeptical of the imperative to matter. And the basis of my skepticism is simple: to argue that something matters is to suggest that the other things, all those trivial dreams/not-dreams that do not have articulate and clever champions like Amira, do not. Personally, I'd like ideally to remain uncertain about, to possibly be astonished by, what does—or ought to—move the world.

Peter Sloterdijk, in *Stress and Freedom*, blames anthropology quite squarely for its role in the broader social-scientific "campaign against amazement," our "resistance to astonishment," the insistence on a "resolutely wonder-free zone" (2016, 2–3). The culprit for him is social constructionism that, as he puts it, "threw the dreamer off the boat," in the figure of an all-too-fleeting Jean-Jacques Rousseau, the dreamer who, as radiant subject, free in his solitary row-row-row-ing—experiencing an "exquisite, ecstatic unusability"—is lassoed back into the world, tasked with the job of *constructing* it. This is society as labor camp: everyone is useful, everyone *makes matter* (Rousseau 1979).

Marilyn Strathern ties anthropology to the adamant dullifying of the universe, too, though her culprit is the rote sensibility of "context." In her 1987 essay "Out of Context: The Persuasive Fictions of Anthropology," Strathern argues that central to the Malinowskian revolution—the break with James Frazer—was the instruction to place alien phenomena within "their context," through the work of ethnography, in order to make those phenomena comparable with "ours"—no longer as hierarchical difference of evolutionary stage but as simply a relative one of differing contexts (Strathern 1987). What might seem amazing to the amateur eye at first—not to worry—is not so amazing after all but is perfectly sensible when professionally placed within the larger context of Trobriand life. Everything, through this narrative, can be mastered.

But can't we ask again, as Strathern did twenty years ago, what is so awful about being amazed? What is all this insistence on the world being ordinary?

Ordinary life, ordinary ethics, ordinary affects. No! If these things are ordinary, it is only because we insist on them being so, making falling chicken feet into a story about political economy or religiosity—a story comparable, rehearsable, *useful*: ethnography as labor camp. But what if we learned a lesson from our dreams (which, according to row-row-row-your-boat philosophy, is also life itself), which obey no single context but cut queerly across them, generating their own context; better yet, generating a unique *situation*? So that's all I want to suggest: that we revisit the philosophical freedom to be and remain astonished, which is enabled first by a skepticism toward context—not throwing it out altogether, which is of course impossible—but refusing to be always beholden to the normative, rote ones (the nation, religion, violence) and instead acknowledging, even generating, the queer contexts; that is—to borrow language from surrealist, antiauthoritarian Marxist movements—the *situations* that allow us to see that everything is extraordinary, to explore the line between saying something and explaining it all away.

And so in what follows, I would like to exercise the pleasurable solidarity of not making things work for me via working too hard to make them work too hard for others. Instead, I want to insist that it's possible to think even of work not in terms of matter and mattering, nor even its dialectical other of ideas and the ideal, but in terms of the useless, the meaningless, the surreal. What possibilities might emerge from this—this refusal of mattering, of making things matter?

One place to begin is with the Situationist International, an antiauthoritarian Marxist group established in France in the late 1950s, with roots in the avant-garde art movements of Dadaism and Surrealism. In their manifesto of 1960, the Situationists anticipate a form of Marxism, exemplified by the Italian autonomists, that rather than valorizing work, finds it indeed boring, dull, and stupid. Their utopian manifesto looks forward to a postwork future made possible by the automation of production and the socialization of essential goods, such that humans will "exude a new surplus value . . . the value of the game, of life freely constructed" ("Situationist Manifesto" 2015, 47). The point is not to do more work or better work, or to be better compensated for work, or to be compensated at all. The vision here is a world of joyful dilettantes, of "amateur-professionals," a world in which everyone is an "artist" insofar as everyone "attains the construction of their own life" (49). I'm drawn toward the Situationists not only for their perspective on work but also for their (I think related) antagonism toward the tyranny of context. The "situation" of Situationism means many things. For one, it means a description of the *situation* before us; that is, of the banality of human existence due to the ostensible imperative of work, and thus our consequent need for the "realization of a better game" (47). But "situation" also refers to an autonomy from existing contexts, such that the world around us determines what we do and do not do, what we value and do not value. The Situationist, by contrast, *creates her own situation*, a product of "the freedom to play," a reflection of our "creative autonomy" (47).

It could be argued that the Situationists don't entirely embrace uselessness and in fact do valorize a kind of work—the work of creativity (a valorizing move Kathi Weeks [2011, 82] is critical of, though not explicitly targeting the Situationists). So let me bring in another champion of uselessness, the sixteenth-century French philosopher Michel de Montaigne. The prolific essayist, at the age of thirty-eight, decided that rather than pursue his growing fame he would retire and dedicate his life to reading, writing, and contemplation; that is, to doing nothing useful for anybody. He inscribed on his bookshelves the following pre-Situationist, pre-autonomist manifesto:

> In the year of Christ 1571, at the age of thirty-eight, on the last day of February, his birthday, Michel de Montaigne, long weary of the servitude of the court and of public employments, while still entire, retired to the bosom of the learned virgins, where in calm and freedom from all cares he will spend what little remains of his life, now more than half run out. If the fates permit, he will complete this abode, this sweet ancestral retreat; and he has consecrated it to his freedom, tranquility, and leisure. (Montaigne 1958, ix–x)

I bring Montaigne up, not out of an insistence on narrative incoherence, but because he has proved useful to me and might prove useful to the project of rethinking work in relation to the non/human. In the midst of his idyll of uselessness, between 1575 and 1578, Montaigne wrote one of his longest essays, the obliquely titled "Apology for Raymond Sebond" (Montaigne 1958). It was in this essay that Montaigne fathered modern radical skepticism (and a radical humility that belies his association with European humanism) with the phrase, "What do I know?" Among the vast evidence Montaigne supplies to demonstrate, indeed, how very little he or anyone else knows is a lengthy section called "Man Is No Better than the Animals" (a section fitted between the similarly excoriating "The Vanity of Man" and "Man's Knowledge Cannot Make Him

Good"). "Man Is No Better than the Animals" contains multiple anecdotes about animal being, the wondrousness as well as the intelligence that exceeds our meager comprehension. After writing of a dog described by Plutarch, who artfully followed stage directions to "play dead," he describes more conventionally laboring animals, some oxen at work: "The oxen that served in the royal gardens of Susa to water them and turn certain great wheels for drawing water, to which there were buckets attached (like many that are to be seen in Languedoc), had been ordered to draw up to one hundred turns a day; they were so accustomed to this number that *it was impossible by any force to make them draw one turn more, and having done their task, they stopped short*" (Montaigne 1958, 340; italics mine). The point Montaigne wants to make with this story is that humans are such dummies that it takes us till our adolescence to be able to count to one hundred. But for me, the point is something different, and in fact, never had it been clearer to me (already sympathetic, of course) that animals "work" until I read this passage.

The point, and my argument, is this: we know that nature works not because it works, or because of *how* it works, but because it *refuses* to work.

Toward Usefulness: Three Justifications

In a move toward usefulness, I'll put forward three reasons for why I believe this is an argument worth making; that is, for why we should focus not on nature working but on nature *refusing* to work.

REJECTING THE WORK ETHIC AND "WORK" ITSELF

First, an argument that focuses on nonhuman refusal to work also refuses in principle the liberal politics of inclusion and recognition (Povinelli 2002). While the act of refusal itself is pleasurable enough, this one also serves the larger purpose of refusing to infinitely expand (and thus make infinitely workable) the central exercise of capitalism, that without which it fails: the exercise of *work*. In making a range of arguments for "how nature works," we can be seen to be participating in an important tradition, exemplified as Weeks lays out, by laborist, antiracist, and feminist movements that all sought in different ways to proclaim their constituents' adherence to, and belonging within, the work ethic. The assertion "we (they) work, too" can be a claim to respect, to sympathy, to mattering—to compensation. "But," as Weeks argues, "all of these demands for inclusion serve at the same time to expand the scope of the work ethic to new groups and new forms of labor, *and to reaffirm its power*" (2011, 68; italics mine).

The problem with reaffirming the power of the work ethic (or of productivism) can be explained, of course, via Max Weber (2002), who detailed its centrality to the rise of capitalism. And the problem many of us have *wanting* to reaffirm it, despite what we learn from Weber, can be partly explained via a dominant and largely fair reading of Marx, in which the critique of alienation presupposes the transhistorical, even metaphysical, value of labor (Baudrillard 1975). Labor in Marx is the very essence of worth (the very essence of mattering), and that which must be liberated. The great critic of capitalism was, oddly, also a productivist.

Weeks agrees, though, unlike some of the harsher autonomist (or *operaismo*) critics like Mario Tronti, she allows that Marx had a little-acknowledged anti-productivist streak, citing for example his assertion in *Capital* that "the reduction of the working day is the basic prerequisite" for freedom (Marx 1981, 959). Tronti, in his 1965 manifesto, a cornerstone of operaismo, "The Strategy of the Refusal," prefers to say that the entire Marxian vision of the worker as the provider of labor is wrong (Tronti 1979). The worker does not provide labor; the *capitalist* provides labor; the worker provides *capital*. The implications of this — as well as the critique of productivism in general, regardless of one's perspective on Tronti's argument — are enormous for thinking about the relationship of nature and work. If without capitalism there is no labor, then labor is obviously not a transhistorical feature of human life, which means (1) that humans have no proprietary relationship to labor (a point we all already agree with) but also (2) that we might be going about this wrong. Instead of taking the position that nature is not mere capital — that it *works* — the more truly liberatory position might be that nature *does* not work, it *should* not work, and neither should we. To return to Weeks, to argue otherwise is ultimately to reaffirm the work ethic. Instead of "nature works, too," we might organize around the uselessness and refusal of work, a rallying point "that can pose the full measure of its antagonism with the exigencies of capital accumulation" (Weeks 2011, 29).

WHAT ANIMALS TEACH US ABOUT POLITICS

The second justification for my argument takes on from the first, in that it rejects the implication of a liberal-paternalist argument of "animals: they're just like us" and instead moves toward the more accurate recognition that human politics would do well to be based more on that of animals. I draw here on Brian Massumi's *What Animals Teach Us about Politics* (2014) and also, again, on Tronti's vision of a strategy of refusal (1979).

As the editors of special issue of *Rethinking Marxism* noted recently (Bloois, Jansen, and Korsten 2014, 164), there has been a surge of interest in the Italian autonomist and operaismo movements that goes sometimes under the label of "post-autonomia." But as the editors also note—and had been my hunch—the post-autonomists seem to invert one of the most fundamental and radical premises of autonomism, which is the inherent power of those who work—specifically, their power to *refuse*. Instead, as we variously see within the anthropology of neoliberalism (where autonomist references are prevalent), the focus tends to be on dire precarity, powerlessness, and need. I'm not suggesting that precarity, powerlessness, and need are not real phenomena, and ones related to the exigencies of capitalism. But I am suggesting that autonomism's refusal-to-work offers other visions and strategies and, furthermore, that it does far more to unsettle humanism than much of the contemporary anthropology of neoliberalism does, which as I see it, recuperates the most anthropocentric and humanist tendencies in the Marxist tradition.

One of the things I like about Tronti is that he has no time for humanism or, for that matter, traditions of any sort. In fact, he puts it thusly: "Nothing [is] more repugnant than the concept of tradition" (1979, 13). The only continuity worth respecting, he goes on to say, is that of the struggle against the imperative of work. And one of the traditions he seeks to unsettle—more so I'd say than some of his colleagues, like Antonio Negri—is assuming the importance of language (that is, of speech, negotiation, and articulation) in political activity. For example, Tronti offers two divergent models of worker revolt: one, a ten-thousand-person march and contract negotiation in 1842, and the other, a day in 1871 on which prisoners who had participated in the Paris Commune but *also* in the June Days Uprising two decades before were ordered separated from the others, shot, and killed. The temptation, Tronti says, is to interpret the first as an offensive action on the part of workers and the second as an act of repression on the part of capitalists; however, "perhaps it is quite the opposite" (1979, 14). His argument is that the first is ultimately a positive development for capitalism (it improves it; it adds consent and spectacle) while the second is "a 'No'" that refuses to "manage" or "improve" society as it stands and therefore must be "repressed by pure violence" (14). While the same might not be true for Tronti

(I can't say), I have no interest here in romanticizing death as a political strategy. I'd rather assume neither life nor death nor work have any inherent mattering. But what I am interested in pointing out via Tronti are two things. First, that speaking and concerted action are not only unnecessary for radical politics but might also be antithetical to it. In other words, that doing nothing (refusing) is more radical than doing something (improving or expanding work). And secondly, again with Weeks, by drawing on a refusal-to-work paradigm, and therefore rejecting rather than growing the reach of the work ethic, we bring back centrally into our analyses the question of *domination*, of, as Tronti puts it, "pure violence." The problem again with the Marxian alienation thesis is its assumption that work and its capacity are already present in living beings (for humanists, in "humans," and for posthumanists, in nature too). The autonomist (and Weeks's) refusal of work, by rejecting that assumption, suggests instead that living beings must be *dominated* in order to work (see also Tronti 1979, 10–11), thus recalling "a focus on . . . power and authority, relations of rulers and the ruled" (Weeks 2011, 20–21). Taking force seriously as the prime mover of all work might help us recognize a real, as opposed to paternalistic, anti/posthumanist commonality.

Dream/Not-Dream

I'll admit straight away that the following was not a dream (whatever that means!). Though I have had many dreams about it. In May 2013, I was living in Hyderabad, spending many of my days with Jayasimha, an activist from the Humane Society International (India). With some controversy, the city of Hyderabad, with the backing of major players in the meat industry, had constructed a large mechanized slaughterhouse on the outskirts of the city (now in Telangana state) in a village called Chengicherla. The objective was to close down the illegal (but de facto acceptable) small butcher shops in the city as well as the scattered legal *mandis*, or live markets, and force butchers as well as sellers and consumers of fresh meat to conduct their business at Chengicherla. A similar move had been made years earlier in Delhi, closing smaller mandis and forcibly relocating live sales and slaughter to a behemoth market on the outskirts of the city in a village called Gazipur. In both cases, the justification for the move was better hygiene, transport, and mechanization (animal welfare was even invoked), though of course the real reason was the smoother accumulation of capital. Nothing proved smooth about Chengicherla, though. The mechanized abattoir was built, three stories and imposing, but within no more than a few months lay vacant, windows shattered and derelict. Did the butchers refuse to come? No, not exactly. They came but refused to use the abattoir. It was inefficient, they said, and the labyrinthine process meant that they could never be certain that the carcasses they received were from the animals they had brought. Instead, they began using the holding pens—one for goats and other small animals and one for cows and buffalo—as collective slaughtering grounds.

Needless to say, the holding pens are not very large. They were designed as intermediary spaces, between arrival and processing, and needless to say, too, they did not come equipped with mechanized tools of slaughter or with its associated logic of concealment and visual opacity (Pachirat 2011). This was a free-for-all, an honest to goodness hell.

Jayasimha brought me there one Sunday. He picked me up at 4:30 a.m. so that we'd arrive by 6:00 a.m. and could see the hundreds of trucks being unloaded. We parked the car. The smell was strong, of bones being burned, of animal and sweat and shit (not manure). But it was the sound that was stunning, like approaching a stadium while a game is in progress, the muted, disembodied din of fifty thousand screaming beings. But where we stood there were bunches of goats waiting around, rather placidly, two-wheelers going past with live animals they had just bought or bags full of meat. A man in a dhoti threw a stick with all his might at a goat for no discernible reason; it was just standing there.

Jayasimha led us to where the trucks were unloading. The trucks were over a story tall. A man would grab a goat by the ear or sometimes a leg and let it hover momentarily (where it seemed to try, with its hooves, to hold on to something), before dropping it, where it either bounced first on the lip of the truck or fell straight down to the ground, several feet lower. Scattered around were downed goats, ones with unnaturally bent limbs, left in the sun to die or, more likely, be dragged to the killing pens with greater force. Most had to be dragged anyway, even if they weren't visibly injured. They were usually pulled by one back leg, the animal trying to gain balance on the front two legs to keep from being dragged along the stony dirt path, but never to any avail. Most cried out in pain and anger and distress, one in particular who simply refused to go but had no choice in the matter in the end. He was dragged mightily, and finally another man had to come and help—or, rather, take over—who just pulled it up and along by the ear. The goat stopped for a moment while it tried to register this new sensation but then continued to groan and scream. We trailed behind, toward the pen, as the stadium sound grew louder.

One of the things I found extraordinary about this is that the man who took over had been following behind the goat as it resisted, mocking its cries. "Bah," "wah," and so on, chuckling to himself. It wasn't his cruelty or lack of compassion that surprised me; quite the contrary. It was that the goat registered *that much* to him that he was compelled to imitate it. Belied here, I realized, was his *sentimentality* about the goat, his *projective identification* with it.

Massumi talks quite a bit about imitation in his text on learning a vital, nonconforming politics from animals (2014, 82–87). He rejects the anthropocentric obsession with the so-called problem of anthropomorphism, arguing that the act of imagining oneself in another, or another in oneself, is a ludic, inventive gesture, a form of *sympathy* (rather than sentimentality) that refuses to be contained or governed categorically. That said, Massumi also recognizes that some identifications are vital ones (such as that of a child's with animality—ludic, inventive play, to return to the Situationists) while some are sterile ones (such as the man's *imitative* identification with the goat that, instead of being inventive, assumes an "as if," an already given set of forms in which it is the animal's place to be dragged and the man's, observing, to be amused and to pretend "as if" he were the animal). But, as I suggested, there is far more to the sterile identification—that is, to the imitation—than an expression of hierarchy. The man's ostensible identificatory indifference ("that is its place, and this is mine") serves, as Massumi puts it, "as a medium for conveying sameness of form" (82). Notice, Massumi says, that animals never imitate humans (they do not make "as if" categories exist or are reversed); *only humans imitate animals*. And so why would an adult human imitate an animal? Because we seek, however poorly, to be inventive and not rote. We seek to stamp ourselves with an "animal motif" (82).

And so here is what I want to suggest about the working man and the working goat who refused. (1) To refuse is inventive (the fewer words used the better). (2) We identify with the inventive refusal. And that is what animals teach us about politics.

REVERSE ARGUMENTATION (KAMADHENU'S LAST STAND)

When we want to say that something isn't "work," we try to show that it is done instinctively, naturally, *generously*. In other words, we know that something isn't "really" work when (we convince ourselves) it issues forth with abundance or internal, instinctive compulsion: a mother's love, a wife's cooking, a flight attendant's smile, a nanny's affection, an intellectual's thinking, a bee's hive making, an artist's art, a cow's milk. So it only makes sense to me that if one wants to argue that something *is* work, one must show that it is *refused*.

Dream within a Dream / Not-Dream

When I saw retired brigadier general S. S. Chauhan in Gurgaon, I was taken with his gentle appearance, his pleasant, open face, his white mustache, his corduroy pants and pageboy cap, and sneakers caked with grass and manure. He was more like a boy than any image I had of a retired brigadier general. But the boyishness belied an ardent passion: the man was *mad* about cows. His wife thinks of the cows as his mistresses. But to him—always the boy—the cows are his mother. "We have an obligation to accept her gift," he said to me, just as we have an obligation to accept the gift of life and nurturance our mothers provide. But I'm a vegan, and while I hate debating politics, I had to raise some objections. He placed his face in his hands, to hide the shock of such a nice girl saying something so stupid. "Why would you not drink your mother's [that is, the *cow's*] milk? That is *why she is here*! What else on earth is she *going to do*?"[1] The true cruelty to the *gau mata* is to refuse her gift, which is a spiritual infraction but also a physical one, leaving her heavy and in pain, bearing her burden for the thankless.

I found him one morning again at his *gaushala*, his cow shelter, and while most are sorry, overcrowded affairs, Chauhan's place was different. Nagar Nigam Gaushala was like a paradise: green grass, visiting schoolchildren dancing under a gazebo, devotional music playing over loudspeakers for the benefit of the cows, around 1,200, all with expanses of space to roam, dropping wholesome dung, waste from their healthy, meticulously balanced diet. Chauhan led me around the grounds, quizzing me on his favorite breeds to see what I had retained from my previous visit. I guessed a Gir correctly, with a prominent hump, high forehead, and foot-long ears that lope downward. Then I spotted another Gir heifer who was passionately mounting another, with all the associated sounds of ecstasy. Chauhan gently steered me away: "Yes, that's the 'unnatural cow.' She believes she's a bull and even when she gets pregnant, she miscarries."[2]

Ah, but then he spotted his favorite cow, Kamadhenu, who would never refuse anything. He ran toward her, asking me over and over again as he held her face in his hands, kissing her and stroking her dewlap, isn't she beautiful? I didn't know about that, but, "What a name!" I said. The giver of all wishes! The name is literally wish cow: *kama* (desire) plus *dhenu* (cow). Chauhan bestowed the name on her because, like the original Kamadhenu, she always had milk to give and could grant any other desire too. "Go on," he said, "whisper a wish in her ear." I did. Chauhan beamed.

And then I remembered a story, the Kamadhenu story of the peaceably self-milking cow, the font of fulfillment, whose divinity allows her to tell a godly person or an innocent from a *kassai* (butcher); she will give abundantly to the former but fall down dead in front of the latter rather than struggle or be killed. Legend has it that she was cursed by Indra to live on earth as an ordinary cow, and in order to be relieved of this curse—for who wants to be ordinary!—carried water from the Ganges in her ear to perform a rite (*abhishek*) on the Siva lingam. Siva, in a fit of pique, threatened to kill this intrusive ordinary cow, but Kamadhenu, selflessly, asked only that he allow her to complete her puja of him first. It was this abundantly self-giving nature of the gau mata, embodied most perfectly by Kamadhenu the actual and the mythical, that Chauhan returned to in order to explain his consumption of milk, his breeding of purebred cows for *panchgavya* (or "cowpathy"), and his selling milk from the gaushala to locals. And because of the company he kept as a member of the Animal Welfare Board, which included vegan activists, he *was* asked to explain himself often. But always his response was the same: "Why would you not drink your mother's milk? That is *why she is here*! What else is she *going to do*?"

But then a funny thing happened. One morning, between 4:00 and 5:00 a.m.—the time of night that Chauhan tends to receive communication from his cows through dreams—he dreamed of Kamadhenu. She came to him and said, "I have been mistreated. I was left out in the cold. A worker abused me." That day he looked for her immediately and asked the workers what had happened, if any of this was true. One admitted that Kamadhenu had refused to give milk lately, and maybe he had been a bit rough with her udders.

And from that day on, retired brigadier general Chauhan gave up milk, and curd, and the sale of anything acquired from his cows.

Coda: Row, Row, Row Your Boat

When I bring up "row, row, row your boat," which I do often, people commonly remark on one of the ironies of the rhyme: why would one row downstream?

The stream is doing all the work, so one should just sit there and relax. Surely then the rhyme is wrong, and life is much more than a dream. But then I am reminded of the radiant subject, Rousseau, alone on his lake, row-row-row-ing in his rapturous, idle dream state, and then I think: not all work deserves to be resisted. Just the useful, mattering, kind. The kind that insists the world is ordinary.

Notes

1. S. S. Chauhan, personal communication, Gurgaon, India, June 21, 2011.
2. S. S. Chauhan, personal communication, Gurgaon, India, February 13, 2013.

Acurio, Gastón. 2015. *Peru: The Cookbook*. London: Phaidon Press.

———. 2016. *Sazón en acción: Algunas recetas para el Perú que queremos*. Lima, Peru: Mitin.

Albuquerque, José L. C. 2005. "Campesinos Paraguayos y "Brasiguayos" en la frontera este del Paraguay." In *Enclave sojero: Merma de soberanía y pobreza*, edited by Ramón B. Fogel and Marcial Riquelme, 149–82. Asunción, Paraguay: Centro de Estudios Rurales Interdisciplinarios.

Alexander, Susan J., David Pilz, Nancy S. Weber, Ed Brown, and Victoria A. Rockwell. 2002. "Mushrooms, Trees, and Money: Value Estimates of Commercial Mushrooms and Timber in the Pacific Northwest." *Environmental Management* 30 (1): 129–41.

Alfrink, Kars, Irene van Peer, and Hein Lagerweij. 2012. "Pig Chase." Playing with Pigs (website). Accessed January 21, 2019. http://www.playingwithpigs.nl.

Andrews, Thomas G. 2008. *Killing for Coal: America's Deadliest Labor War*. Cambridge, MA: Harvard University Press.

Appel, Hannah. 2012. "Offshore Work: Oil, Modularity, and the How of Capitalism in Equatorial Guinea." *American Ethnologist* 39 (4): 692–709.

———. 2014. "Occupy Wall Street and the Economic Imagination." *Cultural Anthropology* 29 (4): 602–25.

Archetti, Eduardo P. 1997. *Guinea Pigs: Food, Symbol and Conflict of Knowledge in Ecuador*. Translated by Valentina Napolitano. New York: Bloomsbury.

Arendt, Hannah. 1998. *The Human Condition*. 2nd ed. Chicago: University of Chicago Press.

———. 2002. "Karl Marx and the Tradition of Western Political Thought." *Social Research* 69 (2): 273–319.

Arens, Richard, ed. 1976. *Genocide in Paraguay*. Philadelphia, PA: Temple University Press.

Arioka, Toshiyuki. 1997. *Matsutake*. Tokyo: Hosei University Press.

Arora, David. 2008. "The Houses That Matsutake Built." *Economic Botany* 62 (3): 278–90.

Atanasoski, Neda, and Kalindi Vora. 2015. "Surrogate Humanity: Posthuman Networks and the (Racialized) Obsolesce of Labor." *Catalyst: Feminism, Theory, Technoscience* 1 (1). https://doi.org/10.28968/cftt.v1i1.28809.

Baildon, Samuel. 1882. *The Tea Industry in India: A Review of Finance and Labor and a Guide for Capitalists*. London: W. H. Allen.

Barad, Karen. 2007. *Meeting the Universe Halfway: Physics and the Entanglement of Matter and Meaning*. Durham, NC: Duke University Press.

Barbiers, Robyn B. 1985. "Orangutans' Color Preference for Food Items." *Zoo Biology* 4 (3): 287–90.

Barnes, Jessica, and Michael R. Dove, eds. 2015. *Climate Cultures: Anthropological Perspectives on Climate Change*. New Haven, CT: Yale University Press.

Bartley, Emma. "In Pictures: 10 Dishes to Try before You Die." *National* (United Arab Emirates), November 14, 2013. https://www.thenational.ae/lifestyle/food/in-pictures-10-dishes-to-try-before-you-die-1.463094#5.

Barua, Maan. 2017. "Nonhuman Labour, Encounter Value, Spectacular Accumulation: The Geographies of a Lively Commodity." *Transactions of the Institute of British Geographers* 42 (2): 274–88.

Battistoni, Alyssa. 2017. "Bringing in the Work of Nature: From Natural Capital to Hybrid Labor." *Political Theory* 45 (1): 5–31.

Baudrillard, Jean. 1975. *The Mirror of Production*. Translated by Mark Poster. St. Louis, MO: Telos.

Bear, Laura. 2014. "Doubt, Conflict, Mediation: The Anthropology of Modern Time." *Journal of the Royal Anthropological Institute* 20:3–30. https://doi.org/10.1111/1467-9655.12091.

Bear, Laura, Karen Ho, Anna Tsing, and Sylvia Yanagisako. 2015. "Gens: A Feminist Manifesto for the Study of Capitalism." Theorizing the Contemporary, *Cultural Anthropology* website, March 30, 2015. http://www.culanth.org/fieldsights/650-generating-capitalism.

Beldo, Les. 2017. "Metabolic Labor: Broiler Chickens and the Explosion of Vitality." *Environmental Humanities* 9 (1): 108–28.

Benezra, Amber, Joseph DeStefano, and Jeffrey I. Gordon. 2012. "Anthropology of Microbes." *Proceedings of the National Academy of Sciences of the United States of America* 109 (17): 6378–81. https://doi.org/10.1073/pnas.1200515109.

Ben-Jacob, Eshel, Alin Finkelshtein, Gil Ariel, and Colin Ingham. 2016. "Multispecies Swarms of Social Microorganisms as Moving Ecosystems." *Trends in Microbiology* 24 (4): 257–69. https://doi.org/10.1016/j.tim.2015.12.008.

Berry, Albert. 2010. "The Role of Agriculture." In *Losing Ground in the Employment Challenge: The Case of Paraguay*, edited by Albert Berry, 61–84. New Brunswick, NJ: Transaction.

Berry, R. Albert, and William R. Cline. 1979. *Agrarian Structure and Productivity in Developing Countries*. Baltimore, MD: Johns Hopkins University Press.

Bertoni, Moisés. 1922. *La civilización guaraní*. Puerto Bertoni, Alto Paraná, Paraguay: Imprenta y edición "Ex Silvis."

———. 1926. *El rozado sin quemar: Una gran reforma necesaria y urgente*. Puerto Bertoni, Alto Paraná, Paraguay: Imprenta y edición "Ex Silvis."

———. 1927a. *El algodón y los algodoneros*. Puerto Bertoni, Alto Paraná, Paraguay: Imprenta y edición "Ex Silvis."

———. 1927b. "La Vida." *Revista de la Sociedad Científica del Paraguay* 2 (3): 139–46.

Besky, Sarah. 2014. *The Darjeeling Distinction: Labor and Justice on Fair-Trade Plantations in India*. Berkeley: University of California Press.

———. 2017. "Fixity: On the Inheritance and Maintenance of Tea Plantation Houses in Darjeeling, India." *American Ethnologist* 44 (4): 617–31.

Bishop, Holley. 2005. *Robbing the Bees: A Biography of Honey—The Sweet Liquid Gold That Seduced the World.* New York: Free Press.

Blanc, Jacob. 2015. "Enclaves of Inequality: Brasiguaios and the Transformation of the Brazil-Paraguay Borderlands." *Journal of Peasant Studies* 42 (1): 145–58.

Blanchette, Alex. 2015. "Herding Species: Biosecurity, Posthuman Labor, and the American Industrial Pig." *Cultural Anthropology* 30 (4): 640–69.

———. 2018. "How to Act Industrial around Industrial Pigs." In *Living with Animals: Bonds across Species*, edited by Natalie Porter and Ilana Gershon, 130–41. Ithaca, NY: Cornell University Press.

———. 2019. "Infinite Proliferation, or The Making of the Modern Runt." *Life by Algorithms: How Roboprocesses Are Remaking Our World*, edited by Catherine Besteman and Hugh Gusterson, 91–106. Chicago: University of Chicago Press.

Blasius, Wilhelm. 1976. *Problems of Life Research: Physiological Analyses and Phenomenological Interpretations.* Berlin: Springer-Verlag.

Bloois, Joost de, Monica Jansen, and Frans Willem Korsten. 2014. "Introduction: From Autonomism to Post-Autonomia, from Class Composition to a New Political Anthropology?" *Rethinking Marxism* 26 (2): 163–77.

Braun, Bruce. 2002. *The Intemperate Rainforest: Nature, Culture, and Power on Canada's West Coast.* Minneapolis: University of Minnesota Press.

Braverman, Harry. 1998. *Labor and Monopoly Capital: The Degradation of Work in the Twentieth Century.* 25th anniversary ed. New York: Monthly Review Press.

Brooks, Daniel, and Michael McClean. 2012. *Summary Report: Boston University Investigation of Chronic Kidney Disease in Western Nicaragua, 2009–2012.* Boston University School of Public Health report to the Compliance Advisor Ombudsman. http://www.cao-ombudsman.org/cases/document-links/documents/BU_SummaryReport_August122012.pdf.

Brooks, Daniel, Oriana Ramirez-Rubio, and Juan Jose Amador. 2012. "CKD in Central America: A Hot Issue." *American Journal of Kidney Diseases* 59 (4): 481–84.

Brown, Wendy. 1995. *States of Injury: Power and Freedom in Late Modernity.* Princeton, NJ: Princeton University Press.

Campbell-Smith, Gail, Miran Campbell-Smith, Ian Singleton, and Matthew Linkie. 2011. "Raiders of the Lost Bark: Orangutan Foraging Strategies in a Degraded Landscape." *PLoS ONE* 6 (6): e20962. https://doi.org/10.1371/journal.pone.0020962.

Cánepa, Gisela. 2013. "Nation Branding: The Re-foundation of Community, Citizenship, and the State in the Context of Neoliberalism in Perú." *Medien Journal* 37 (3): 7–18.

Carlson, Alvar W. 1986. "Ginseng: America's Botanical Drug Connection to the Orient." *Economic Botany* 40 (2): 233–49.

Carse, Ashley. 2014. *Beyond the Big Ditch: Politics, Ecology, and Infrastructure at the Panama Canal.* Cambridge, MA: MIT Press.

Carter, Michael R. 1984. "Identification of the Inverse Relationship between Farm Size and Productivity: An Empirical Analysis of Peasant Agricultural Production." *Oxford Economic Papers* 36 (1): 131–45.

Castañeda, Claudia. 2002. *Figurations: Child, Bodies, Worlds*. Durham, NC: Duke University Press.

Caton, Steven C. 2005. *Yemen Chronicle: An Anthropology of War and Mediation*. New York: Hill and Wang.

Chamney, Montfort. 1930. *The Story of the Tea Leaf*. Calcutta: New India Press.

Chatterjee, Piya. 2001. *A Time for Tea: Women, Labor, and Post/Colonial Politics on an Indian Plantation*. Durham, NC: Duke University Press.

Chatterjee, Rhitu. 2016. "Mysterious Kidney Disease Goes Global." *Science*, March 31, 2016. https://www.sciencemag.org/news/2016/03/mysterious-kidney-disease-goes-global.

Choi, Vivian. 2015. "Anticipatory States: Tsunami, War, and Insecurity in Sri Lanka." *Cultural Anthropology* 30 (2): 286–309.

Choy, Tim, and Jerry Zee. 2015. "Condition: Suspension." *Cultural Anthropology* 30 (2): 210–23.

Collier, Russell, and Tom Hobby. 2010. "It's All about Relationships: First Nations and Non-timber Resource Management in British Columbia." *BC Journal of Ecosystems and Management* 11 (1–2): 1–8.

Collins, Jane L. 2003. *Threads: Gender, Labor, and Power in the Global Apparel Industry*. Chicago: University of Chicago Press.

———. 2017. *The Politics of Value: Three Movements to Change How We Think about the Economy*. Chicago: University of Chicago Press.

Collins, Jane L., and Martha Gimenez, eds. 1990. *Work without Wages: Comparative Studies of Domestic Labor and Self-Employment*. Albany: SUNY Press.

Colloredo-Mansfeld, Rudi. 1998. "'Dirty Indians,' Radical Indígenas, and the Political Economy of Social Difference in Modern Ecuador." *Bulletin of Latin American Research* 17 (2): 185–205.

Comaroff, Jean. 1985. *Body of Power, Spirit of Resistance: The Culture and History of a South African People*. Chicago: University of Chicago Press.

Coulter, Kendra. 2016. *Animals, Work, and the Promise of Interspecies Solidarity*. New York: Palgrave Macmillan.

Crate, Susan A., and Mark Nuttall. 2009. *Anthropology and Climate Change: From Encounters to Actions*. Walnut Creek, CA: Left Coast Press.

Cronon, William. 1995. "The Trouble with Wilderness." In *Uncommon Ground: Rethinking the Human Place in Nature*, edited by William Cronon, 69–90. New York: W. W. Norton.

Crowley-Matoka, Megan, and Sherine Hamdy. 2016. "Gendering the Gift of Life: Family Politics and Kidney Donation in Egypt and Mexico." *Medical Anthropology* 35 (1): 31–44.

CVR (La Comisión de la Verdad y Reconciliación). 2004. *Hatun Willakuy: Versión abreviada del informe final de la Comisión de la Verdad y Reconciliación, Perú*. Lima, Peru: Corporación Gráfica.

Debnath, Sailen. 2010. *The Dooars in Historical Transition*. Siliguri, West Bengal: N. L. Publishers.

Degregori, Carlos Ivan. 2012. *How Difficult It Is to Be God: Shining Path's Politics of War in Peru, 1980–1999*. Madison: University of Wisconsin Press.

de la Cadena, Marisol. 2000. *Indigenous Mestizos: The Politics of Race and Culture in Cuzco, Peru, 1919–1991*. Durham, NC: Duke University Press.

———. 2010. "Indigenous Cosmopolitics in the Andes: Conceptual Reflections beyond 'Politics.'" *Cultural Anthropology* 25 (2): 334–70.

De León, Jason. 2015. *The Land of Open Graves: Living and Dying on the Migrant Trail*. Oakland: University of California Press.

Derrida, Jacques. 1981. "The Pharmakon." In *Dissemination*. Translated by Barbara Johnson. Chicago: University of Chicago Press.

———. 1982. *Margins of Philosophy*. Translated by Alan Bass. Chicago: University of Chicago Press.

Descola, Philippe, and Gisli Palsson, eds. 1996. *Nature and Society: Anthropological Perspective*. London: Routledge.

Dey, Arnab. 2015. "Bugs in the Garden: Tea Plantations and Environmental Constraints in Eastern India (Assam), 1840–1910." *Environment and History* 21 (4): 537–65.

di Leonardo, Micaela. 1987. "The Female World of Cards and Holidays: Women, Families, and the Work of Kinship." *Signs: Journal of Women in Culture and Society* 12 (3): 440–53.

DiNovelli-Lang, Danielle, and Karen Hébert. 2018. "Ecological Labor." Theorizing the Contemporary, *Cultural Anthropology* website, July 26, 2018. https://culanth.org/fieldsights/1510-ecological-labor.

Dominus, Susan. 2010. "The Mystery of the Red Bees of Red Hook." *New York Times*, November 29, 2010. https://www.nytimes.com/2010/11/30/nyregion/30bigcity.html.

Douglas, Mary. 1966. *Purity and Danger: An Analysis of Concepts of Pollution and Taboo*. New York: Routledge.

Dudley, Kathryn Marie. 1994. *The End of the Line: Lost Jobs, New Lives in Postindustrial America*. Berkeley: University of California Press.

Durkheim, Emile. 1997. *The Division of Labor in Society*. Translated by W. D. Halls. New York: Free Press.

Ehlers, Tyson, Signy Fredrickson, and Shannon Berch. 2007. "Pine Mushroom Habitat Characteristics and Management Strategies in the West Kootenay Region of British Columbia." *Journal of Ecosystems and Management* 8 (3): 76–88.

Ehrenreich, Barbara. 2001. *Nickel and Dimed: On (Not) Getting By in America*. New York: Metropolitan Books.

Elgert, Laureen. 2016. "'More Soy on Fewer Farms' in Paraguay: Challenging Neoliberal Agriculture's Claims to Sustainability." *Journal of Peasant Studies* 43 (2): 537–61.

Elias, Norbert. 1994. *The Civilizing Process: Sociogenetic and Psychogenetic Investigations*. Translated by Edmund Jephcott. Oxford: Blackwell.

Ellen, Roy. 1999. "Modes of Subsistence and Ethnobiological Knowledge: Between Extraction and Cultivation in Southeast Asia." In *Folkbiology*, edited by Douglas L. Medin and Scott Atran, 91–117. Cambridge, MA: MIT Press.

Engels, Frederich. 2009. *The Condition of the Working Class in England*. London: Penguin.

Esposito, Roberto. 2008. *Bíos: Biopolitics and Philosophy*. Translated by Timothy C. Campbell. Minneapolis: University of Minnesota Press.

Fabian, Johannes. 2002. *Time and the Other: How Anthropology Makes Its Object*. New York: Columbia University Press.

Fan, Judith. 2013. "Can Ideas about Food Inspire Real Social Change? The Case of Peruvian Gastronomy." *Gastronomica* 13 (2): 29–40.

Farmer, Paul. 1999. *Infections and Inequalities: The Modern Plagues*. Berkeley: University of California Press.

Federici, Silvia. 1975. *Wages Against Housework*. Montpelier, Bristol, UK: Power of Women Collective/Falling Wall Press.

———. 2018. *Re-enchanting the World: Feminism and the Politics of the Commons*. Oakland, CA: PM Press.

Fennell, Catherine. 2015. *Last Project Standing: Civics and Sympathy in Post-Welfare Chicago*. Minneapolis: University of Minnesota Press.

Ferguson, James. 2015. *Give a Man a Fish: Reflections on the New Politics of Distribution*. Durham, NC: Duke University Press.

Fogel, Ramón B. 1998. *Mbyá recové: La resistencia de un pueblo indómito*. Asunción, Paraguay: Centro de Estudios Rurales Interdisciplinarios.

———. 2005. "Efectos socioambientales del enclave sojero." In *Enclave sojero: Merma de soberanía y pobreza*, edited by Ramón B. Fogel and Marcial Riquelme, 35–112. Asunción, Paraguay: Centro de Estudios Rurales Interdisciplinarios.

Fogel, Ramón B., and Marcial Riquelme, eds.. 2005. *Enclave sojero: Merma de soberanía y pobreza*. Asunción, Paraguay: Centro de Estudios Rurales Interdisciplinarios.

Forestry and Forest Products Research Institute, Mushroom Research Laboratory. 2008. "Matsutake no DAN Gensankoku Hanbetsuho" (DNA identification method of the country of origin of matsutake mushrooms). Accessed on May 3, 2018. http://www.ffpri.affrc.go.jp/labs/matsutake/#asia_matsutake.

Fortun, Kim. 2012. "Ethnography in Late Industrialism." *Cultural Anthropology* 27 (3): 446–64.

Fortun, Mike. 2008. *Promising Genomics: Iceland and deCODE Genetics in a World of Speculation*. Berkeley: University of California Press.

Foster, John Bellamy. 1999. "Marx's Theory of Metabolic Rift: Classical Foundations for Environmental Sociology." *American Journal of Sociology* 105 (2): 366–405.

Foster, John Bellamy, Brett Clark, and Richard York. 2010. *The Ecological Rift: Capitalism's War on the Earth*. New York: Monthly Review Press.

Foucault, Michel. 1990. *The History of Sexuality, Vol. 1: An Introduction.* New York: Vintage Books.

Franklin, Sarah, and Margaret Lock. 2003. *Remaking Life and Death: Toward an Anthropology of the Biosciences.* Santa Fe, NM: SAR Press.

Franklin, Sarah, and Susan McKinnon. 2001. *Relative Values: Reconfiguring Kinship Studies.* Durham, NC: Duke University Press.

Fraser, David. 2008. *Understanding Animal Welfare: The Science in Its Cultural Context.* London: Wiley-Blackwell.

Fraser, Laura. 2006. "Next Stop Lima." *Gourmet,* August 2006. http://www.gourmet. com.s3-website-us-east-1.amazonaws.com/magazine/2000s/2006/08/ nextstoplima.html.

Freeman, Derek. 1955. *Iban Agriculture: A Report on the Shifting Cultivation of Hill Rice by the Iban of Sarawak.* London: Her Majesty's Stationary Office.

Frutos, Juan Manuel. 1976. *De la reforma agraria al bienestar rural y otros documentos concernientes a la marcha de la reforma agraria.* Asunción, Paraguay: Instituto de Bienestar Rural.

Galdikas, Biruté. 1988. "Orangutan Diet, Range, and Activity at Tanjung Puting, Central Borneo." *International Journal of Primatology* 9 (1): 1–35.

García, María Elena. 2005. *Making Indigenous Citizens: Identity, Development, and Multicultural Activism in Peru.* Stanford, CA: Stanford University Press.

———. 2013. "The Taste of Conquest: Colonialism, Cosmopolitics, and the Dark Side of Peru's Gastronomic Boom." *Journal of Latin American and Caribbean Anthropology* 18 (3): 505–24.

Gibson-Graham, J. K. 2006. *The End of Capitalism (As We Knew It): A Feminist Critique of Political Economy.* Minneapolis: University of Minnesota Press.

Gieryn, Thomas F. 1983. "Boundary-Work and the Demarcation of Science from Non-Science: Strains and Interests in Professional Ideologies of Scientists." *American Sociological Review* 48 (6): 781–95.

Gilliam, Martha. 1997. "Identification and Roles of Non-Pathogenic Microflora Associated with Honey Bees." *FEMS Microbiology Letters* 155 (1): 1–10. https://doi. org/10.1111/j.1574-6968.1997.tb12678.x.

Glaucer, Marcos. 2009. *Extranjerización del territorio paraguayo.* Asunción, Paraguay: Base Investigaciones Sociales.

Gordon, Deborah. 2016. "The Queen Does Not Rule." *Aeon,* December 19, 2016. https://aeon.co/essays/how-ant-societies-point-to-radical-possibilities-for-humans.

Gorz, André. 1994. *Capitalism, Socialism, Ecology.* Translated by Martin Chalmers. New York: Verso.

Gould, Jeffrey L. 1990. *To Lead as Equals: Rural Protest and Political Consciousness in Chinandega, Nicaragua, 1912–1979.* Chapel Hill: University of North Carolina Press.

Graeber, David. 2013. "It Is Value That Brings Universes into Being." *HAU: Journal of Ethnographic Theory* 3 (2): 219–43.

———. 2018. *Bullshit Jobs: A Theory.* New York: Simon and Schuster.

Graeter, Stephanie. 2017. "To Revive an Abundant Life: Catholic Science and Neoextractivist Politics in Peru's Mantaro Valley." *Cultural Anthropology* 32 (1): 117–48.

Grandin, Temple. 2005. *Animals in Translation: Using the Mysteries of Autism to Decode Animal Behavior.* New York: Scribner.

Griffiths, Percival. 1967. *The History of the Indian Tea Industry.* London: Weidenfeld and Nicolson.

Guin, Jerry. 1997. *Matsutake Mushroom: The White Goldrush of the 1990s: A Guide and Journal.* Happy Camp, CA: Naturegraph Publishers.

Guthman, Julie. 2004. *Agrarian Dreams: The Paradox of Organic Farming in California.* Berkeley: University of California Press.

———. 2016. "Going Both Ways: More Chemicals, More Organics, and the Significance of Land in Post–Methyl Bromide Fumigation Decisions for California's Strawberry Industry." *Journal of Rural Studies* 47 (A): 76–84.

Hall, Stuart. 1980. "Race, Articulation, and Societies Structured in Dominance." In *Sociological Theories: Racism and Colonialism,* 305–45. Paris: UNESCO.

Harari, Yuval Noah. 2017. "The Meaning of Life in a World without Work." *Guardian,* May 8, 2017. www.theguardian.com/technology/2017/may/08/virtual-reality-religion-robots-sapiens-book.

Haraway, Donna J. 1991. *Simians, Cyborgs, and Women: The Reinvention of Nature.* New York: Routledge.

———. 1992. "Ecce Homo, Ain't (Ar'n't) I a Woman, and Inappropriate/d Others: The Human in a Post-Humanist Landscape." In *Feminists Theorize the Political,* edited by Judith Butler and Joan W. Scott, 86–100. London: Routledge.

———. 2008. *When Species Meet.* Minneapolis: University of Minnesota Press.

———. 2015. "Anthropocene, Capitalocene, Plantationocene, Cthulucene: Making Kin." *Environmental Humanities* 6 (1): 159–65.

———. 2016. *Staying with the Trouble: Making Kin in the Chthulucene.* Durham, NC: Duke University Press.

Hardt, Michael, and Antonio Negri. 2000. *Empire.* Cambridge, MA: Harvard University Press.

———. 2004. *Multitude: War and Democracy in the Age of Empire.* New York: Penguin Press.

Harris, Marvin. 1966. "The Cultural Ecology of India's Sacred Cattle." *Current Anthropology* 7 (1): 51–66.

Harris, Olivia. 2000. *To Make the Earth Bear Fruit: Essays on Fertility, Work and Gender in Highland Bolivia.* London: Institute for Latin American Studies.

Harrisson, Barbara. 1962. *Orang-Utan.* London: Collins.

Harshey, Rasika M., and Jonathan D. Partridge. 2015. "Shelter in a Swarm." In "Cooperative Behaviour in Microbial Communities," edited by Ian Barry Holland, Nicola R. Stanley-Wall, Sarah J. Coulthurst, and Karin Sauer. Special issue, *Journal of Molecular Biology* 427 (23): 3683–94.

Hartigan, John, Jr. 2013. "Mexican Genomics and the Roots of Racial Thinking." *Cultural Anthropology* 28 (3): 372–95.

———. 2014. *Aesop's Anthropology: A Multispecies Approach*. Minneapolis: University of Minnesota Press.

Harvey, Chuck. 2012. "New Almond Promises Independence from Bees." *Business Journal*, February 9, 2012. http://www.thebusinessjournal.com/news/agriculture/770-new-almond-promises-independence-from-bees.

Hayden, Cori. 2003. *When Nature Goes Public: The Making and Unmaking of Bioprospecting in Mexico*. Princeton, NJ: Princeton University Press.

Helmreich, Stefan. 2008. "Species of Biocapital." *Science as Culture* 17 (4): 463–78.

———. 2009. *Alien Ocean: Anthropological Voyages in Microbial Seas*. Berkeley: University of California Press.

Hemsworth, Paul H. 2003. "Human-Animal Interactions in Livestock Production." *Applied Animal Behaviour Science* 81 (3): 185–98.

Hetherington, Kregg. 2011. *Guerrilla Auditors: The Politics of Transparency in Neoliberal Paraguay*. Durham, NC: Duke University Press.

———. 2013. "Beans before the Law: Knowledge Practices, Responsibility, and the Paraguayan Soy Boom." *Cultural Anthropology* 28 (1): 65–85.

Hinyub, Chris. 2012. "Lower Honey Production Not Necessarily Linked to CCD." *IVN*, April 20, 2012. http://ivn.us/2012/04/20/lower-honey-production-not-necessarily-linked-to-ccd/.

Hochschild, Arlie Russell. 1983. *The Managed Heart: The Commercialization of Human Feeling*. Berkeley: University of California Press.

Hock, William Gwee Thian. 2006. *A Baba Malay Dictionary: The First Comprehensive Compendium of Straits Chinese Terms and Expressions*. Rutland, VT: Tuttle Publishing.

Hodges, Matt. 2014. "Immanent Anthropology: A Comparative Study of 'Process' in Contemporary France." *Journal of the Royal Anthropological Institute* 20 (S1): 33–51.

Hoffman, Gloria D., Russell Vreeland, Diana Sammataro, and Ruben Alarcon Jr. 2009. "The Importance of Microbes in Nutrition and Health of Honey Bee Colonies Part-3: Where Do We Go from Here?" *American Bee Journal* 149:755–57.

Holmes, Seth M. 2013. *Fresh Fruit, Broken Bodies: Migrant Farmworkers in the United States*. Berkeley: University of California Press.

Horton, Sarah Bronwen. 2016. *They Leave Their Kidneys in the Field: Illness, Injury, and Illegality among U.S. Farmworkers*. Oakland: University of California Press.

Hosford, David, David Pilz, Randy Molina, and Michael Amaranthus. 1997. *Ecology and Management of the Commercially Harvested American Matsutake Mushroom*. Portland, OR: US Department of Agriculture, Forest Service.

Hostetler, Chris. "2012 Research Review: Top Reasons Producers Cull Sows." *National Hog Farmer*, March 23, 2012. https://www.nationalhogfarmer.com/reproduction/research-reviews-top-reasons-producers-cull-sows.

Humane Society of the United States. 2010. *Undercover at Smithfield Foods*. December 15, 2010. Video, 3:35. https://www.youtube.com/watch?v=L_vqIGTKuQE.

Huner, Michael. Forthcoming. "How Pedro Quiñonez Lost His Soul." *Journal of Social History*.

Imhoff, Daniel, ed. 2010. *CAFO: The Tragedy of Industrial Animal Factories*. San Rafael, CA: Earth Aware.

Ingold, Tim. 1983. "The Architect and the Bee: Reflections on the Work of Animals and Men." *Man* 18 (1): 1–20.

Jain, Sarah Lochlann. 2006. *Injury: The Politics of Product Design and Safety Law in the United States*. Princeton, NJ: Princeton University Press.

Jasanoff, Shelia. 2007. *Designs on Nature: Science and Democracy in Europe and the United States*. Princeton, NJ: Princeton University Press.

Jayaraman, Saru. 2014. *Behind the Kitchen Door*. Ithaca, NY: Cornell University Press.

Jeter, Clay, dir. *Chef's Table*. Season 3, episode 6, "Virgilio Martínez." Released February 17, 2017, on Netflix. https://www.netflix.com/title/80007945.

Johannsen, Kristin. 2006. *Ginseng Dreams: The Secret World of America's Most Valuable Plant*. Lexington: University Press of Kentucky.

Johnson, Elizabeth R. 2015. "Of Lobsters, Laboratories, and War: Animal Studies and the Temporalities of More-Than-Human Encounters." *Environment and Planning D: Society and Space* 33 (2): 296–313.

Kaur, Amarjit. 2004. *Wage Labour in Southeast Asia since 1840: Globalisation, the International Division of Labour and Labour Transformations*. New York: Palgrave Macmillan.

Kawata, Ko. 2016. "Chugoku ni okeru Tanabata Densetsu no Seishinshi" [Spiritual history of Tanabata legends in China]. *Ningen Bunka Kenkyu* (Kyoto Gakuen Daigaku Ningen Bunkagakkai) 37:149–78.

Kedit, Peter Mulok. 1980. *Modernization among the Iban of Sarawak*. Kuala Lumpur: Dewan Bahasa dan Pustaka, Kementerian Pelajaran Malaysia.

Keegan, William F., and Lisabeth A. Carlson. 2008. *Talking Taino: Caribbean Natural History from a Native Perspective*. Tuscaloosa: University of Alabama Press.

Kennedy, Jean. 2008. "Pacific Bananas: Complex Origins, Multiple Dispersals?" *Asian Perspectives* 47 (1): 75–94.

Kirksey, Eben, and Stefan Helmreich. 2010. "The Emergence of Multispecies Ethnography." *Cultural Anthropology* 25 (4): 545–76.

Klages, Ludwig. 2011. *Rizumu no Honshitsu ni Tsuite* [Vom Wesen des Rhythmus / The Essence of Rhythms]. Translated by Shinichi Hirasawa and Katsuhiro Yoshimasu. Tokyo: Ubusuna Shoin.

Knott, Cheryl. 1998. "Changes in Orangutan Caloric Intake, Energy Balance, and Ketones in Response to Fluctuating Fruit Availability." *International Journal of Primatology* 19 (6): 1061–79.

Kockelman, Paul. 2016. "Grading, Gradients, Degradation, Grace. Part 1: Intensity and Causality." *Hau* 6 (2): 389–423.

Kohn, Eduardo. 2013. *How Forests Think: Towards an Anthropology Beyond the Human.* Berkeley: University of California Press.

Kolbert, Elizabeth. 2016. "Our Automated Future." *New Yorker,* December 19 and 26, 2016. https://www.newyorker.com/magazine/2016/12/19/our-automated-future.

Kosek, Jake. 2010. "Ecologies of Empire: On the New Uses of the Honeybee." *Cultural Anthropology* 25 (4): 650–78.

Kuriyama, Shigehisa. 2017. "The Geography of Ginseng and the Strange Alchemy of Needs." In *The Botany of Empire in the Long Eighteenth Century,* edited by Yota Batsaki, Sarah Burke Cahalan, and Anatole Tchikine, 61–72. Washington, DC: Dumbarton Oaks Research Library and Collection.

Lamont, Michèle, and Virág Molnár. 2002. "The Study of Boundaries in the Social Sciences." *Annual Review of Sociology* 28 (1): 167–95.

Lamoreaux, Janelle. "What if the Environment Is a Person? Lineages of Epigenetic Science in a Toxic China." *Cultural Anthropology* 31 (2): 188–214.

Lamphere, Louise. 1987. *From Working Daughters to Working Mothers: Immigrant Women in a New England Industrial Community.* Ithaca, NY: Cornell University Press.

Landecker, Hannah. 2013. "Postindustrial Metabolism: Fat Knowledge." *Public Culture* 25 (3): 495–522.

Latour, Bruno. 1993. *We Have Never Been Modern.* Translated by Catherine Porter. Cambridge, MA: Harvard University Press.

———. 2004. *Politics of Nature: How to Bring the Sciences into Democracy.* Translated by Catherine Porter. Cambridge, MA: Harvard University Press.

———. 2005. *Reassembling the Social: An Introduction to Actor-Network Theory.* Oxford: Oxford University Press.

Law, John, and Annemarie Mol. 2002. *Complexities: Social Studies of Knowledge Practices.* Durham, NC: Duke University Press.

Laws, Rebecca L. 2015. "Chronic Kidney Disease of Unknown Etiology in Nicaragua: Investigating the Role of Environmental and Occupational Exposures." PhD diss., Boston University School of Public Health.

Lee, Jack Tsen-Ta. n.d. A Dictionary of Singlish and Singapore English (website). Accessed November 21, 2018. http://www.singlishdictionary.com/.

Lefebvre, Henri. (1992) 2013. *Rhythmanalysis: Space, Time and Everyday Life.* Translated by Stuart Elden and Gerald Moore. London: Bloomsbury.

Li, Tania Murray. 2014. *Land's End: Capitalist Relations on an Indigenous Frontier.* Durham, NC: Duke University Press.

Lifsher, Marc. 2012. "Hives for Hire." *Los Angeles Times,* March 3, 2012. http://articles.latimes.com/2012/mar/03/business/la-fi-california-bees-20120304.

Liu, Lydia H. 1995. *Translingual Practice: Literature, National Culture, and Translated Modernity—China, 1900–1937.* Stanford, CA: Stanford University Press.

Lock, Margaret. 2017. "Recovering the Body." *Annual Review of Anthropology* 46:1–14.

Locke, John. 1980. *Second Treatise of Government*. Edited by C. B. Macpherson. Indianapolis, IN: Hackett Publishing Company.

Lorimer, Jamie. 2015. *Wildlife in the Anthropocene: Conservation after Nature*. Minneapolis: University of Minnesota Press.

Lowe, Celia. 2010. "Viral Clouds: Becoming H5N1 in Indonesia." *Cultural Anthropology* 25 (4): 625–49.

Lutgendorf, Philip. 2012. "Making Tea in India: Chai, Capitalism, Culture." *Thesis Eleven* 113 (1): 11–31.

Mankekar, Purnima, and Akhil Gupta. 2016. "Intimate Encounters: Affective Labor in Call Centers." *Positions* 24 (1): 17–43.

Martínez, Virgilio. 2016. *Central*. London: Phaidon.

Martuccelli, Danilo. 2015. *Lima y sus arenas: Poderes sociales y jerarquías culturales*. Lima, Peru: Cauces Editores.

Marx, Karl. 1973. *Grundrisse*. Translated by Martin Nicolaus. New York: Penguin.

———. 1976. *Capital, Volume 1*. Translated by Ben Fowkes. New York: Penguin.

———. 1981. *Capital, Volume 3*. Translated by David Fernbach. London: Penguin.

Masco, Joseph. 2017. "The Crisis in Crisis." *Current Anthropology* 58 (S15): S65–S75.

Massumi, Brian. 2014. *What Animals Teach Us about Politics*. Durham, NC: Duke University Press.

Matos Mar, José. 2004. *Desborde popular y crisis del estado: Veinte años después*. Lima, Peru: Fondo Editorial del Congreso de la República.

Matta, Raúl. 2013. "Valuing Native Eating: The Modern Roots of Peruvian Food Heritage." *Anthropology of Food* S8. http://journals.openedition.org/aof/7361.

———. 2016. "Recipes for Crossing Boundaries: Peruvian Fusion." In *Cooking Technology: Transformations in Culinary Practice in Mexico and Latin America*, edited by Steffan Igor Ayora-Diaz, 139–52. New York: Bloomsbury.

Mavromichalis, Ioannis. 2011. "Resurgence of Milk Replacers." *Pig Progress Online*. June 22, 2011. https://www.pigprogress.net/Breeding/Sow-Feeding/2011/6/Resurgence-of-milk-replacers-PP007492W/.

Mayer, Enrique. 1991. "Peru in Deep Trouble: Mario Vargas Llosa's 'Inquest in the Andes' Reexamined." *Cultural Anthropology* 6 (4): 466–504.

McClintock, Anne. 1995. *Imperial Leather: Race, Gender, and Sexuality in the Colonial Contest*. London: Routledge.

McCoy, Alfred W. 1982. "A Queen Dies Slowly: The Rise and Decline of Iloilo City." In *Philippine Social History: Global Trade and Local Transformations*, edited by Alfred W. McCoy and Ed. C. de Jesus, 271–76. Quezon City, Philippines: Ateneo de Manila University Press.

McDermott Hughes, David. 2017. "A Jobless Utopia?" *Boston Review*, May 19, 2017. http://bostonreview.net/class-inequality/david-mcdermott-hughes-jobless-utopia.

McGowan, Alexander. 1860. *Tea Planting in the Outer Himalayah*. London: Smith, Elder, and Co.

McLaren, Duncan. 2015. "Where's the Justice in Geoengineering?" *Guardian*, March 14,

2015. www.theguardian.com/science/political-science/2015/mar/14/
wheres-the-justice-in-geoengineering.

McWilliams, Carey. 2000. *Factories in the Field: The Story of Migratory Farm Labor in California*. Berkeley: University of California Press.

Méndez G., Cecilia. 1996. "Incas Sí, Indios No: Notes on Peruvian Creole Nationalism and Its Contemporary Crisis." *Journal of Latin American Studies* 28 (1): 197–225.

Menzies, Charles R. 2006. "Ecological Knowledge, Subsistence, and Livelihood Practices: The Case of the Pine Mushroom Harvest in Northeastern British Columbia." In *Traditional Ecological Knowledge and Natural Resource Management*, edited by Charles R. Menzies, 87–104. Lincoln: University of Nebraska Press.

Merchant, Carolyn. 1980. *The Death of Nature: Women, Ecology, and the Scientific Revolution*. San Francisco, CA: HarperSanFrancisco.

Millar, Kathleen. 2014. "The Precarious Present: Wageless Labor and Disrupted Life in Rio de Janeiro, Brazil." *Cultural Anthropology* 29 (1): 32–53.

Miller, Dale. 2007. "Before You Target 30 p/s/y Read This." *National Hog Farmer*, February 15, 2007. https://www.nationalhogfarmer.com/mag/farming_target_psy_read.

Minamino, Tohru, Katsumi Imada, and Keiichi Namba. 2008. "Molecular Motors of the Bacterial Flagella." *Current Opinion in Structural Biology* 18 (6): 693–701.

Mintz, Sidney W. 1960. *Worker in the Cane: A Puerto Rican Life History*. New Haven, CT: Yale University Press.

———. 1979. "Time, Sugar, and Sweetness." *Marxist Perspectives* 2 (4): 56–73.

———. 1985. *Sweetness and Power: The Place of Sugar in Modern History*. New York: W. W. Norton.

Mitchell, Don. 1996. *Lie of the Land: Migrant Workers and the California Landscape*. Minneapolis: University of Minnesota Press.

Mitchell, Timothy. 2002. *Rule of Experts: Egypt, Techno-Politics, Modernity*. Berkeley: University of California Press.

Mittermaier, Amira. 2011. *Dreams that Matter: Egyptian Landscapes of the Imagination*. Berkeley: University of California Press.

Montaigne, Michel de. 1958. *The Complete Essays of Montaigne*. Translated by Donald M. Frame. Palo Alto, CA: Stanford University Press.

Moore, Amelia. 2015. "Islands of Difference: Design, Urbanism, and Sustainable Tourism in the Anthropocene Caribbean." *Journal of Latin American and Caribbean Anthropology* 20 (3): 513–32.

———. 2016. "Anthropocene Anthropology: Reconceptualizing Contemporary Global Change." *Journal of the Royal Anthropological Institute* 22 (1): 27–46.

Moore, Donald S., Jake Kosek, and Anand Pandian, eds. 2003. *Race, Nature, and the Politics of Difference*. Durham, NC: Duke University Press.

Moore, Henrietta L. 1986. *Space, Text, and Gender: An Anthropological Study of the Marakwet of Kenya*. Cambridge: Cambridge University Press.

Moore, Jason W. 2015. *Capitalism in the Web of Life: Ecology and the Accumulation of Capital.* London: Verso Books.

———. 2016. "The Rise of Cheap Nature." In *Anthropocene or Capitalocene? Nature, History, and the Crisis of Capitalism,* edited by Jason W. Moore, 78–115. Oakland, CA: PM Press.

Morales, Edmundo. 1995. *The Guinea Pig: Healing, Food, and Ritual in the Andes.* Tucson: University of Arizona Press.

Morrison, Kathleen. 2015. "Provincializing the Anthropocene." *Seminar* 673:75–80.

Mortimer, Peter E., Samantha C. Karunarathna, Qiaohong Li, Heng Gui, Xueqing Yang, Xuefei Yang, Jun He, et al. 2012. "Prized Edible Asian Mushrooms: Ecology, Conservation and Sustainability." *Fungal Diversity* 56 (1): 31–47.

Moseley-Williams, Sorrel. 2015. "Meet Virgilio Martinez, Peru's Best Chef." CNN, *Culinary Journeys,* October 2015. http://www.cnn.com/2015/10/09/foodanddrink/profile-chef-virgilio-martinez-central-peru-lima/.

Muehlebach, Andrea. 2011. "On Affective Labor in Post-Fordist Italy." *Cultural Anthropology* 26 (1): 59–82.

Muehlebach, Andrea, and Nitzan Shoshan. 2012 "Post-Fordist Affect: An Introduction." *Anthropological Quarterly* 85 (2): 317–43.

Mukharji, Projit Bihari. 2014. "Vishalyakarani as *Eupatorium ayapana*: Retrobotanizing, Embedded Traditions, and Multiple Historicities of Plants in Colonial Bengal, 1890–1940." *Journal of Asian Studies* 73 (1): 65–87.

Münzel, Mark. 1973. *The Aché Indians: Genocide in Paraguay.* IWGIA DOCUMENT Series No. 11. Copenhagen: International Work Group for Indigenous Affairs.

Murphy, Michelle. 2006. *Sick Building Syndrome and the Problem of Uncertainty: Environmental Politics, Technoscience, and Women Workers.* Durham, NC: Duke University Press.

Murray, Douglas L. 1994. *Cultivating Crisis: The Human Cost of Pesticides in Latin America.* Boulder, CO: Westview Press.

Myers, Natasha. 2015. *Rendering Life Molecular: Models, Modelers, and Excitable Matter.* Durham, NC: Duke University Press.

Nading, Alex M. 2015a. "Chimeric Globalism: Global Health in the Shadow of the Dengue Vaccine." *American Ethnologist* 42 (2): 356–70.

———. 2015b. *Mosquito Trails: Ecology, Health, and the Politics of Entanglement.* Oakland: University of California Press.

NAOJ (National Astronomical Observatory of Japan). n.d. "Yoku Aru Shitsumon 3–9: Tanabata ni tsuite Oshiete" [Frequently asked questions 3–9: Tanabata story]. Accessed May 4, 2018. https://www.nao.ac.jp/faq/a0309.html.last.

Nash, June. 1979. *We Eat the Mines and the Mines Eat Us: Dependency and Exploitation in Bolivian Tin Mins.* New York: Columbia University Press.

———. 1994. "Global Integration and Subsistence Insecurity." *American Anthropologist* 96 (1): 7–30.

Negri, Antonio. 1989. *The Politics of Subversion: A Manifesto for the Twenty-First Century.* Translated by James Newell. Cambridge: Polity Press.

Ngidang, Dimbab. 2003. "Transformation of the Iban Land Use System in Post Independence Sarawak." *Borneo Research Bulletin* 34:62–78.

Nordhaus, Hannah. 2011. *The Beekeeper's Lament: How One Man and Half a Billion Honey Bees Help Feed America*. New York: Harper Perennial.

Nugent, Guillermo. 2012. *El laberinto de la choledad*. Lima, Peru: Universidad Peruana de Ciencias Aplicadas.

Ogawa, Makoto. 1978. *Matsutake no Seibutsugaku* [Biology of Matsutake Mushroom]. Tokyo: Tsukiji Shokan.

Ong, Aihwa. 1987. *Spirits of Resistance and Capitalist Discipline: Factory Women in Malaysia*. Albany: SUNY Press.

———. 1991. "The Gender and Labor Politics of Postmodernity." *Annual Review of Anthropology* 20:279–309.

———. 2003. *Buddha Is Hiding: Refugees, Citizenship, and the New America*. Berkeley: University of California Press.

Pachirat, Timothy. 2011. *Every Twelve Seconds: Industrialized Slaughter and the Politics of Sight*. New Haven, CT: Yale University Press.

Parreñas, [Rheana] Juno Salazar. 2012. "Producing Affect: Transnational Volunteerism in a Malaysian Orangutan Rehabilitation Center." *American Ethnologist* 39 (4): 673–87.

———. 2017. "Orangutan Rehabilitation as an Experiment of Decolonization." In "Critical Perspectives: Engagements with Decolonization and Decoloniality in and at the Interfaces of STS," curated and introduced by Kristina Lyons, Juno Parreñas, and Noah Tamarkin. Special issue, *Catalyst: Feminism, Theory, Technoscience* 3 (1): 19–25. https://catalystjournal.org/index.php/catalyst/article/view/28794.

———. 2018. *Decolonizing Extinction: The Work of Care in Orangutan Rehabilitation*. Durham, NC: Duke University Press.

Partridge, Jonathan D., and Rasika M. Harshey. 2013. "Swarming: Flexible Roaming Plans." *Journal of Bacteriology* 195 (5): 909–18.

PASE/ILRF. 2005. *Labor Conditions in the Nicaraguan Sugar Industry*. Report submitted to and distributed by the International Labor Rights Fund. http://digitalcommons.ilr.cornell.edu/cgi/viewcontent.cgi?article=2018&context=globaldocs.

Paxson, Heather. 2010. "Locating Value in Artisanal Cheese: Reverse Engineering *Terroir* for New-World Landscapes." *American Anthropologist* 112 (3): 444–57.

———. 2018. "The Naturalization of Nature as Working." Theorizing the Contemporary, *Cultural Anthropology* website, July 26, 2018. https://culanth.org/fieldsights/1508-the-naturalization-of-nature-as-working.

Perez, Patricia, dir. 2011. *Mistura: The Power of Food*. Lima, Peru: Chiwake Films. DVD.

Pilz, David, and Randy Molina, eds. 1994. *Managing Forest Ecosystems to Conserve Fungus Diversity and Sustain Wild Mushroom Harvest*. Portland, OR: US Department of Agriculture, Forest Service.

Plumwood, Val. 2008. "Shadow Places and the Politics of Dwelling." *Australian*

Humanities Review 44:139–50. http://australianhumanitiesreview.
org/2008/03/01/shadow-places-and-the-politics-of-dwelling/.

Pollan, Michael. 2001. *The Botany of Desire: A Plant's-Eye View of the World*. New York: Random House.

Porcher, Jocelyne. 2015. "Animal Work." In *The Oxford Handbook of Animal Studies*, edited by Linda Kalof, 302–18. Oxford: Oxford University Press.

Porter, Natalie. 2013. "Bird Flu Biopower: Strategies for Multispecies Coexistence in Việt Nam." *American Ethnologist* 40 (1): 132–48.

Postone, Moishe. 1993. *Time, Labor, and Social Domination: A Reinterpretation of Marx's Critical Theory*. Cambridge: Cambridge University Press.

Povinelli, Elizabeth A. 2002. *The Cunning of Recognition: Indigenous Alterities and the Making of Australian Multiculturalism*. Durham, NC: Duke University Press.

———. 2011. *Economies of Abandonment: Social Belonging and Endurance in Late Liberalism*. Durham, NC: Duke University Press.

Rabinbach, Anson. 1992. *The Human Motor: Energy, Fatigue, and the Origins of Modernity*. Berkeley: University of California Press.

Raffles, Hugh. 2010. *Insectopedia*. New York: Pantheon.

Ramirez-Rubio, Octavia, Michael D. McLean, Juan José Amador, and Daniel Brooks. 2013. "An Epidemic of Chronic Kidney Disease in Central America: An Overview." *Journal of Epidemiology and Community Health* 67:1–3.

Rappaport, Roy A. 1968. *Pigs for the Ancestors: Ritual in the Ecology of a New Guinea People*. Long Grove, IL: Waveland Press.

Richards, Peter D. 2011. "Soy, Cotton, and the Final Atlantic Forest Frontier." *Professional Geographer* 63 (3): 343–63.

Rico Numbela, Elizabeth. 2010. Conozca al centro "MEJOCUY" de Bolivia (website). http://ricardo.bizhat.com/rmr-prigeds/cadena-del-cuy.htm.

Rifkin, Jeremy. 1995. *The End of Work: The Decline of the Global Labor Force and the Dawn of the Post-Market Era*. New York: Tarcher/Putnam.

Rioja-Lang, Fiona C., Yolande M. Seddon, and Jennifer A. Brown. 2018. "Shoulder Lesions in Sows: A Review of Their Causes, Prevention, and Treatment." *Journal of Swine Health and Production* 26 (2): 101–7.

Robin, Marie-Monique. 2010. *The World According to Monsanto: Pollution, Corruption, and the Control of the World's Food Supply*. Translated by George Holoch. New York: New Press.

Røder, Henriette L., Søren J. Sørensen, and Mette Burmølle. 2016. "Studying Bacterial Multispecies Biofilms: Where to Start?" *Trends in Microbiology* 24 (6): 503–13.

Rodman, Peter S., and John G. H. Cant. 1984. *Adaptations for Foraging in Nonhuman Primates: Contributions to an Organismal Biology of Prosimians, Monkeys, and Apes*. New York: Columbia University Press.

Rodríguez, Luis Aliaga, Roberto Moncayo Galliano, Elizabeth Rico, and Alberto Caycedo. 2009. *Producción de cuyes*. Lima, Peru: Fondo Editorial UCSS.

Rojas Villagra, Luis. 2015. *Campesino rape: Apuntes teóricos e históricos sobre el campesinado y la tierra en Paraguay*. Asunción, Paraguay: Base Investigaciones Sociales.

Roncal-Jimenez, Carlos, Ramón García-Trabanino, Lars Barregard, Miguel A. Lanaspa, Catharina Wesseling, Tamara Harra, Aurora Aragón, et al. 2016. "Heat Stress Nephropathy from Exercise-Induced Uric Acid Crystalluria: A Perspective on Mesoamerican Nephropathy." *American Journal of Kidney Diseases* 67 (1): 20–30.

Rosaldo, Renato. 1986. "Ilongot Hunting as Story and Experience." In *The Anthropology of Experience*, edited by Victor T. Turner and Edward M. Brunner, 97–138. Urbana: University of Illinois Press.

Rose, Deborah Bird. 2004. *Reports from a Wild Country: Ethics for Decolonisation*. Sydney, Australia: University of New South Wales Press.

———. 2013. *Wild Dog Dreaming: Love and Extinction*. Charlottesville: University of Virginia Press.

Rousseau, Jean-Jacques. 1979. *The Reveries of the Solitary Walker*. Translated by Charles E. Butterworth. New York: New York University Press.

Russell, Edmund. 2004. "The Garden in the Machine: Toward an Evolutionary History of Technology." In *Industrializing Organisms: Introducing Evolutionary History*, edited by Susan R. Schrepfer and Philip Scranton, 1–18. London: Routledge.

Sahlins, Marshall. 1972. *Stone Age Economics*. Chicago: Aldine-Atherton.

Saito, Haruo, and Gaku Mitsumata. 2008. "Bidding Customs and Habitat Improvement for Matsutake (*Tricholoma matsutake*) in Japan." *Economic Botany* 62 (3): 257–68.

Saldaña-Portillo, María Josefina. 2003. *The Revolutionary Imagination in the Americas and the Age of Development*. Durham, NC: Duke University Press.

Salzinger, Leslie. 2003. *Genders in Production: Making Workers in Mexico's Global Factories*. Berkeley: University of California Press.

Santos, Jesús M., dir. 2012. *Peru Sabe: Cuisine as an Agent of Social Change*. Lima, Peru: Media Networks and Tensacalma Productions, 2013. DVD.

Satsuka, Shiho. 2015. *Nature in Translation: Japanese Tourism Encounters the Canadian Rockies*. Durham, NC: Duke University Press.

———. 2019. "Translation in the World Multiple." In *The World Multiple: The Quotidian Politics of Knowing and Generating Entangled Worlds*, edited by Keiichi Omura, Grant Jun Otsuki, Shiho Satsuka, and Atsuro Morita, 219–32. Abingdon, UK: Routledge.

Schabas, Margaret. 2003. "Adam Smith's Debts to Nature." *History of Political Economy* 35 (5): 262–81.

Schmehl, Daniel R., Peter E. A. Teal, James L. Frazier, and Christina M. Grozinger. 2014. "Genomic Analysis of the Interaction between Pesticide Exposure and Nutrition in Honey Bees (*Apis mellifera*)." *Journal of Insect Physiology* 71:177–90. https://doi.org/10.1016/j.jinsphys.2014.10.002.

Schneider, Andrew. 2011. "Asian Honey, Banned in Europe, Is Flooding U.S. Grocery

Shelves." *Food Safety News*, August 15, 2011. http://www.foodsafetynews.
com/2011/08/honey-laundering/.

Schneider, Mindi, and Phillip McMichael. 2010. "Deepening, and Repairing, the Metabolic Rift." *Journal of Peasant Studies* 37 (3): 461–84.

Scott, James C. 1976. *The Moral Economy of the Peasant: Rebellion and Resistance in
Southeast Asia*. New Haven, CT: Yale University Press.

Segal, Daniel A., and Sylvia J. Yanagisako. 2005. *Unwrapping the Sacred Bundle: Reflections on the Disciplining of Anthropology*. Durham, NC: Duke University
Press.

Shapiro, Nicholas. 2015. "Attuning to the Chemosphere: Domestic Formaldehyde,
Bodily Reasoning, and the Chemical Sublime." *Cultural Anthropology* 30 (3):
368–93.

Sharma, Jayeeta. 2011. *Empire's Garden: Assam and the Making of Modern India*. Durham, NC: Duke University Press.

Shillington, Laura. 2013. "Right to Food, Right to the City: Household Urban Agriculture and Socio-Natural Metabolism in Managua, Nicaragua." *Geoforum* 44
(1): 103–11.

"Situationist Manifesto." 2015. Translated by Fabian Tompsett. In *Cosmonauts of the
Future: Texts from the Situationist Movement in Scandinavia and Elsewhere*,
edited by Mikkel Bolt and Jacob Jakobsen, 47–49. Copenhagen: Nebula.

Sloterdijk, Peter. 2016. *Stress and Freedom*. Malden, MA: Polity Press.

Smith, Neil. 2008. *Uneven Development: Nature, Capital, and the Production of Space*.
Athens: University of Georgia Press.

Soper, Kate. 1995. *What Is Nature?* Oxford: Blackwell Publishers.

Souchaud, Sylvain. 2002. *Pionniers brésiliens au Paraguay*. Paris: Karthala.

Srnicek, Nick, and Alex Williams. 2015. *Inventing the Future: Postcapitalism and a
World without Work*. New York: Verso.

Stengers, Isabelle. 2010. *Cosmopolitics I*. Translated by Robert Bononno. Minneapolis:
University of Minnesota Press.

Stern, Steve J., ed. 1998. *Shining and Other Paths: War and Society in Peru, 1980–1995*.
Durham, NC: Duke University Press.

Stoler, Ann Laura. 1995. *Race and the Education of Desire: Foucault's History of Sexuality and the Colonial Order of Things*. Durham, NC: Duke University Press.

Strathern, Marilyn. 1980. "No Nature, No Culture: The Hagen Case." In *Nature, Culture
and Gender*, edited by Carol P. MacCormack and Marilyn Strathern, 174–222.
Cambridge: Cambridge University Press.

———. 1987. "Out of Context: The Persuasive Fictions of Anthropology" [and comments and reply]. *Current Anthropology* 28 (3): 251–81.

———. 1988. *The Gender of the Gift: Problems with Women and Problems with Society
in Melanesia*. Berkeley: University of California Press.

———. 1992. *After Nature: English Kinship in the Late Twentieth Century*. Cambridge:
Cambridge University Press.

Suh, Soyoung. 2008. "Herbs of Our Own Kingdom: Layers of the 'Local' in the *Materia Medica* of Early Choson Korea." *Asian Medicine* 4 (2): 395–422.

Sunder Rajan, Kaushik. 2006. *Biocapital: The Constitution of Postgenomic Life*. Durham, NC: Duke University Press.

———. 2012. *Lively Capital: Biotechnologies, Ethics, and Governance in Global Markets*. Durham, NC: Duke University Press.

Suzuki, Kazuo. 2005. "Ectomycorrhizal Ecophysiology and the Puzzle of Tricholoma Matsutake." [In Japanese.] *Journal of the Japanese Forest Society* 87 (1): 90–102.

Swettenham, Frank Athelstane. 1887. *Vocabulary of the English and Malay Languages, with Notes*. London: W. B. Whittingham.

Tadiar, Neferti. 2012. "Life-Times in Fate Playing." *South Atlantic Quarterly* 111 (4): 783–802.

Tanaka, Martin, 1998. *Los espejismos de la democracia: El colapso del sistema de partidos en el Peru, 1980–1995, en perspectiva comparada*. Lima, Peru: Instituto de Estudios Peruanos.

Taussig, Michael T. 1980. *The Devil and Commodity Fetishism in South America*. Chapel Hill: University of North Carolina Press.

Tegel, Simeon. 2012. "Peru's Fantastic Food Revolution." *Guardian*, September 21, 2012. https://www.theguardian.com/travel/2012/sep/21/peru-lima-food-restaurants-revolution.

Thacker, Eugene. 2007. "Thought Creatures." *Theory, Culture & Society* 24 (7–8): 314–16.

———. 2010. *After Life*. Chicago: University of Chicago Press.

Thompson, Charis. 2005. *Making Parents: The Ontological Choreography of Reproductive Technologies*. Cambridge, MA: MIT Press.

Thompson, E. P. 1967. "Time, Work-Discipline, and Industrial Capitalism." *Past and Present* 38:56–97.

Toledo, Ricardo. 2010. "Farm Size-Productivity Relationships in Paraguay's Agricultural Sector." In *Losing Ground in the Employment Challenge: The Case of Paraguay*, edited by Albert Berry, 85–100. New Brunswick, NJ: Transaction.

Tracy, Megan. 2013. "Pasteurizing China's Grasslands and Sealing *Terroir*." *American Anthropologist* 115 (3): 437–51.

Tronti, Mario. 1979. "The Strategy of the Refusal." In *Working Class Autonomy and the Crisis: Italian Marxist Texts of the Theory and Practice of a Class Movement*, 7–21. London: Red Notes and CSE Books.

Trubek, Amy, Kolleen M. Guy, and Sarah Bowen. 2010. "Terroir: A French Conversation with a Transnational Future." *Contemporary French and Francophone Studies* 14 (2): 139–48.

Tsai, Yen-Ling, Isabelle Carbonell, Joelle Chevrier, and Anna Lowenhaupt Tsing. 2016. "Golden Snail Opera: The More-Than-Human Performance of Friendly Farming on Taiwan's Lanyang Plain." *Cultural Anthropology* 31 (4): 520–44.

Tsing, Anna Lowenhaupt. 2009. "Supply Chains and the Human Condition." *Rethinking Marxism* 21 (2): 148–76.

———. 2012. "Unruly Edges: Mushrooms as Companion Species." *Environmental Humanities* 1:141–54.

———. 2015. *The Mushroom at the End of the World: The Possibility of Life in Capitalist Ruins*. Princeton, NJ: Princeton University Press.

———. 2019. "A Multispecies Ontological Turn?" In *The World Multiple: The Quotidian Politics of Knowing and Generating Entangled Worlds*, edited by Keiichi Omura, Grant Jun Otsuki, Shiho Satsuka, and Atsuro Morita, 233–47. Abingdon, UK: Routledge.

Uexküll, Jakob von. 2010. *A Foray into the Worlds of Animals and Humans: With A Theory of Meaning*. Translated by Joseph D. O'Neil. Minneapolis: University of Minnesota Press.

USDA (US Department of Agriculture). 1940. *Census of Agriculture*. Vol. 3, *General Report*. http://agcensus.mannlib.cornell.edu/AgCensus/censusParts.do?year=1940.

USDA NRCS (US Department of Agriculture, Natural Resources Conservation Service). 2014. State Agricultural Economic Information Sources (website). http://www.nrcs.usda.gov/wps/portal/nrcs/detailfull/national/technical/econ/costs/?cid=nrcs143_022015.

Valderrama, Mariano. 2016. *¿Cuál es el futuro de la gastronomía peruana?* Lima, Peru: Apega.

VanBuren, Robert, Fanchang Zeng, Cuixia Chen, Jisen Zhang, Ching Man Wai, Jennifer Han, Rishi Aryal, et al. 2015. "Origin and Domestication of Papaya Y^h Chromosome." *Genome Research* 25 (4): 524–33.

Van Dooren, Thom. 2014. *Flight Ways: Life and Loss at the Edge of Extinction*. New York: Columbia University Press.

Van Gestel, Jordi, Hera Vlamakis, and Roberto Kolter. 2015. "From Cell Differentiation to Cell Collectives: *Bacillus subtilis* Uses Division of Labor to Migrate." *PLoS Biology* 13 (4): e1002141. https://doi.org/10.1371/journal.pbio.1002141.

Verisk Maplecroft. 2015. "Heat Stress Threatens to Cut Labour Productivity in SE Asia by up to 25% within 30 years—Verisk Maplecroft." Verisk Maplecroft website, October 15, 2015. Accessed January 15, 2019. https://www.maplecroft.com/portfolio/new-analysis/2015/10/28/heat-stress-threatens-cut-labour-productivity-se-asia-25-within-30-years-verisk-maplecroft/.

Virno, Paolo. 2004. *A Grammar of the Multitude: For an Analysis of Contemporary Forms of Life*. Translated by Isabella Bertoletti, James Cascaito, and Andrea Casson. New York: Semiotext(e).

Viveiros de Castro, Eduardo. 1998. "Cosmological Deixis and Amerindian Perspectivism." *Journal of the Royal Anthropological Institute* 4 (3): 469–88.

Vogel, Steven. 1988. "Marx and Alienation from Nature." *Social Theory and Practice* 14 (3): 367–87.

Wadiwell, Dinesh. 2018. "Chicken Harvesting Machine: Animal Labor, Resistance, and the Time of Production." *South Atlantic Quarterly* 117 (3): 527–49.

Wagner, Carlos. 1990. *Brasiguaios: Homens sem pátria*. Petrópolis, Brazil: Editora Vozes.

Wallace, Robert G., and Richard A. Kock. 2012. "Whose Food Footprint? Capitalism, Agriculture, and the Environment." *Human Geography* 5 (1): 63–83.

Walley, Christine J. 2013. *Exit Zero: Family and Class in Post-Industrial Chicago*. Chicago: University of Chicago Press.

Weber, Max. 2002. *The Protestant Ethic and the Spirit of Capitalism*. Edited and translated by Peter Baeher and Gordon C. Wells. New York: Penguin.

Weeks, Kathi. 2011. *The Problem with Work: Feminism, Marxism, Antiwork Politics, and Postwork Imaginaries*. Durham, NC: Duke University Press.

Weiss, Brad. 2016. *Real Pigs: Shifting Values in the Field of Local Pork*. Durham, NC: Duke University Press.

West, Paige. 2012. *From Modern Production to Imagined Primitive: The Social World of Coffee from Papua New Guinea*. Durham, NC: Duke University Press.

West, Stuart A., Stephen P. Diggle, Angus Buckling, Andy Gardner, and Ashleigh S. Griffin. 2007. "The Social Lives of Microbes." *Annual Review of Ecology, Evolution, and Systematics* 38 (1): 53–77.

Weszkalnys, Gisa. 2015. "Geology, Potentiality, Speculation: On the Indeterminacy of First Oil." *Cultural Anthropology* 30 (4): 611–39.

White, Richard. 1995a. "'Are You an Environmentalist or Do You Work for a Living?' Work and Nature." In *Uncommon Ground: Rethinking the Human Place in Nature*, edited by William Cronon, 171–85. New York: W. W. Norton.

———. 1995b. *The Organic Machine: The Remaking of the Columbia River*. New York: Hill and Wang.

Wilderson, Frank, III. 2003. "Gramsci's Black Marx: Whither the Slave in Civil Society?" *Social Identities* 9 (2): 225–40.

Wilkinson, Richard James. 1908. *An Abridged Malay-English Dictionary*. Kuala Lumpur, Malaysia: F.M.S. Government Press.

Williams, Raymond. 1976. *Keywords: A Vocabulary of Culture and Society*. New York: Oxford University Press.

Wolf, Eric R. 2001. *Pathways of Power: Building an Anthropology of the Modern World*. Berkeley: University of California Press.

Wolfe, Cary. 2009. *What Is Posthumanism?* Minneapolis: University of Minnesota Press.

Wolfe, Patrick. 2016. *Traces of History: Elementary Structures of Race*. London: Verso.

Woodruff, Mandi. 2012. "One Third of America's Honey May Be a Dangerous and Illegal Import from China." *Business Insider*, February 24, 2012. http://www.businessinsider.com/one-third-of-honey-in-the-us-may-be-an-illegal-and-dangerous-import-from-china-2012-2.

Wu, Yan. 2003. "Jidai no Ehnsen ni tomonau Shinwa Densetsu no Henyo: Tanabata

Densetsu o megutte" [Transformation of myths and legends across historical eras: About tanabata legend]. *Doshisha Kokubungaku* 58:115–23.

Wynter, Sylvia. 2003. "Unsettling the Coloniality of Being/Power/Truth/Freedom: Towards the Human, After Man, Its Overrepresentation—An Argument." *CR: The New Centennial Review* 3 (3): 257–337.

Yamamoto, Dorothy. 2016. *Guinea Pig*. New York: Reaktion Books.

Yanagisako, Sylvia Junko. 2002. *Producing Culture and Capital: Family Firms in Italy*. Princeton, NJ: Princeton University Press.

———. 2012. "Immaterial and Industrial Labor: On False Binaries in Hardt and Negri's Trilogy." *Focaal* 64:16–23.

Yeh, Emily. 2000. "Forest Claims, Conflicts and Commodification: The Political Ecology of Tibetan Mushroom-Harvesting Villages in Yunnan Province, China." *China Quarterly* 161:264–78.

Zavella, Patricia. 1987a. "'Abnormal Intimacy': The Varying Work Networks of Chicana Cannery Workers." *Feminist Studies* 11 (3): 541–57.

———. 1987b. *Women's Work and Chicano Families: Cannery Workers of the Santa Clara Valley*. Ithaca, NY: Cornell University Press.

Zeiderman, Austin. 2016. *The Endangered City: The Politics of Security and Risk in Bogotá*. Durham, NC: Duke University Press.

Participants in the School for Advanced Research Advanced Seminar "How Nature Works," co-chaired by Sarah Besky, Alex Blanchette, and Naisargi N. Dave, September 25–29, 2016. *Front row, from left*: Jake Kosek, Alex Nading, Sarah Besky, Eleana Kim, Thomas G. Andrews. *Middle row, from left*: John Hartigan, Juno Salazar Parreñas, Naisargi N. Dave. *Back row, from left*: Shiho Satsuka, María Elena García, Alex Blanchette, Kregg Hetherington. Photograph by Garret Vreeland. © School for Advanced Research.

THOMAS G. ANDREWS
Department of History, University of Colorado, Boulder

SARAH BESKY
Department of Anthropology and Watson Institute for International and
Public Affairs, Brown University

ALEX BLANCHETTE
Department of Anthropology and Environmental Studies, Tufts University

NAISARGI N. DAVE
Department of Anthropology, University of Toronto

MARÍA ELENA GARCÍA
Comparative History of Ideas Program, Jackson School of International
Studies, University of Washington

JOHN HARTIGAN JR.
Department of Anthropology, University of Texas, Austin

KREGG HETHERINGTON
Department of Sociology and Anthropology, Concordia University

ELEANA KIM
Department of Anthropology, University of California, Irvine

JAKE KOSEK
Department of Geography, University of California, Berkeley

ALEX NADING
Watson Institute for International and Public Affairs, Brown University

JUNO SALAZAR PARREÑAS
Department of Gender and Women's Studies, Ohio State University

SHIHO SATSUKA
Department of Anthropology, University of Toronto

Page numbers in italic text indicate illustrations. Page numbers in bold text indicate authors' contributions to this volume.

Acevedo, Juan, 134
Aché people (Paraguay), 53
activism and resistance: activist exposés of industrial livestock operations, 61, 62; in India, 39; in Nicaragua, 101–2, 104, 106–8, 109–11; in Paraguay, 41, 42, 48, 55–56
Acurio, Gastón, 133, 140, 143
agricultural extension programs, 165
agriculture: bees as essential pollinators, 151–52; the Green Revolution, 42, 44–45, 52, 58; as mark of human settlement, 27; unintentional cultivation of matsutake, 194. *See also* farmworkers; industrial agriculture; *specific countries and crops*
Ahn Seong-gi, 124
alienation, and monocultures, 25, 26, 27, 33–34
Alien Ocean: Anthropological Voyages in Microbial Seas (Helmreich), 184
almond industry (California), 151–53, 156, 158, 161. *See also* bees and beekeeping
American hog farms, 16, 59–76; the animals' work and behavior, 60–63, 64–65, 67–68, 73–74; bedsores and skin workers, 59, 60–61, 76n1; confinement and its impacts, 60–61, 67–68, 75, 76n1; development and features of, 63–64; ethical questions, 74–75; exposés and activism, 61, 62; hog selection and specialization, 65, 69, 74; hormone administration and its impacts, 71–73; human-hog relations on, 60, 62, 63, 64–69, 73, 74, 75;

human labor on, 59, 60, 61–62, 65–68, 69–73, 74
Anderson, Jeff, 150, 156–61, 162, 164, 166
Andrews, Thomas G., vii–viii, 81, 102
animals and animal labor. *See* human-animal relations; multispecies *entries*; nonhuman agency and productivity; *specific animals and settings*
anthropocentrism, 14. *See also* nonhuman agency and productivity
anthropomorphism, 221
anti-productivism, 9, 217
Arendt, Hannah, 118–19, 184–85, 198
Arguedas, José María, 132
Ariel, Gil, 178–79
ASOCHIVIDA (Chichigalpa Association for Life), 107–8, 109–11
Astrid y Gastón (restaurant), 144
attunement, 14; contrasting temporalities and, 203, 206–8; mushroom pickers' sensory attunement, 18, 191–93, 199–201, 206, 207–8
automation. *See* mechanization
autonomism, 218
autonomy: and the *changa* in Paraguay, 51; from ethnographic context, 212–15

bacteria. *See* microbes and microbial behaviors
Bako National Park (Sarawak), 85
Barad, Karen, 113
Batu Wildlife Center, 89
Bear, Laura, 205
bees and beekeeping, 1, 17–18, 149–68; bee care and feeding, 150, 157–63; bee health and population declines, 150,

bees and beekeeping (*continued*)
156, 160–61, 164–66; bees as migrant
laborers/moving of hives, 150, 151–52,
153, 155, 156, 158–59, 160, 163–64,
166; bees as quintessential laborers,
153–54, 167–68; bees' vulnerability
to pesticides, 150, 161, 163, 164–65;
and the California almond industry,
151–53, 156, 158, 161; changes in the
beekeeping industry, 157, 158–60,
165–66; and the decline of agricultural extension programs, 165;
economics of beekeeping, 151, 152,
155–57, 159–60, 161, 166; economic
value of honeybee labor, 152, 156; the
honey industry, 155–57; industrial
agriculture's dependence on, 149, 150,
151–53; Marx on bees, 50, 164; pollination as bees' most needed service,
156–57; urban beekeeping, 162
Benezra, Amber, 184, 185–86
Bertoni, Hernando, 44–45, 52
Bertoni, Moises Santiago, 43–44, 51, 57
Besky, Sarah, 16, **23–40**, 49, 54, 125
biocapital, 13
biofilms, 172, 177, 186, 187, 188
biotechnology, 13–14; genetically engineered crops, 42, 53, 54, 55–56, 163
Blanchette, Alex, 16, **59–76**
boundaries and boundary work,
188–89; microbial border crossers,
181–82, 189
Braverman, Harry, 50, 62–63
Brazilian migrants, as soy farmers in
Paraguay, 41, 47–48
British Columbia, matsutake picking in.
See matsutake mushroom picking
Bungo Segu, 85
Burmølle, Mette, 188

California agriculture: almonds and
bees, 151–53, 156, 158, 161. *See also* bees
and beekeeping

Canada, matsutake picking in. *See*
matsutake mushroom picking
Capital (Marx). *See* Marx, Karl
capitalist ruins, 25–26, 27
Capitalocene, 119
cari makan (finding food) metaphor,
79–81, 83, 87–88, 92, 93
Castañeda, Claudia, 136, 144
CCZ (Civilian Control Zone), South
Korea, 115, 116–17, 119, 127. *See also*
Korean ginseng
changas (day wages) and *changueros*,
48–49, 51–52, 54–55
Chauca, Lilia, 131–32
Chauhan, S. S., 222–23
chemical use: on American hog farms,
71–73; bees' vulnerability to, 150, 161,
163, 164–65; and the economics of
industrial agriculture, 154; in Korean
ginseng farming, 116, 124, 125–26; in
Nicaraguan agriculture, 106, 107; in
Paraguayan agriculture, 42, 52–54,
55–56, 57–58; Roundup and Roundup
Ready crops, 42, 53, 54, 55–56, 163
Chichigalpa Association for Life
(ASOCHIVIDA), 107–8, 109–11
Ching (orangutan), 88, 90–91, 92–93
chronic kidney disease (CKD) in
Nicaragua: among sugarcane workers,
97, 100–102, 104, 106–8, 109–11; city
dwellers' experiences of, 98–100, 102,
108–9; CKDnt form, 97, 100–102, 106,
107–8, 109–11, 113; the role of heat in,
97–98, 109–11, 113
ChungKwanJang (CKJ) ginseng, 122,
124–25
Civilian Control Zone (CCZ), South
Korea, 115, 116–17, 119, 127. *See also*
Korean ginseng
CKD. *See* chronic kidney disease
CKJ (ChungKwanJang) ginseng, 122,
124–25
Clark, Brett, 118

climate change, 108, 110, 126. *See also* environmental instabilities; heat; Nicaraguan heatscapes/workscapes

coffee, 33

Colloredo-Mansfeld, Rudi, 141–42

commodity fetishism, 124, 167, 174, 189n3

communication: mushroom pickers' sensory attunement as, 193, 199–201, 206–8; nonhuman work as communicative practice, 197–98

communities: of nonhumans, 183. *See also* sociality; solidarity

context, tyranny of, in ethnography, 212–15

cotton, 33; in Nicaragua, 106; in Paraguay, 42–46, 48–49, 51, 53–54, 58, 58n2

cranes, impact of ginseng farming on, 115–17, 129

creativity, 7, 214–15

CTC tea production and consumption, 24, 30–31

cuy (guinea pig). *See* guinea pig

Cylindrocarpon destructans, 125–26

Darwin, Charles, 182, 183

Dave, Naisargi N., 18, **211–24**

deforestation, in Paraguay, 42, 53, 54, 56, 57

demilitarized zone (Korea). *See* DMZ

Derrida, Jacques, 128, 129

DeStefano, Joseph, 184, 185–86

diabetes, 97, 98, 99–100, 104

disappearing root rot, 125–26

division of labor: and animality, 65; by microbes, 177–79, 182, 183

DMZ and DMZ ginseng, 116, 119–20, 126, 127, 129. *See also* Korean ginseng

DMZ EcoResearch, 115, 128

Douglas, Mary, 141

Dreams That Matter (Mittermaier), 213

Duncans Industries Limited, 24

Durkheim, Emile, 177, 178

eco-technology, 5

El algodón y los algodoneros (Cotton and the Cotton Growers) (Bertoni), 43, 44

El Cuy (comic strip character), 134

Elias, Norbert, 142

Ellen, Roy, 83

Ellis, Steve, 153, 158, 159, 161, 162, 164, 165, 166

end of work. *See* postwork futures and imaginaries

endurance, 27, 33–34, 38, 40

environmental instabilities, 2–3, 4–7, 9–10; the ecological or metabolic rift notion, 10, 17, 118

ethnography, 15. *See also* multispecies ethnography

exhaustion/fatigue, 33–34, 36, 102; of monocultures, 6–7, 25–26, 27, 36; in Nicaraguan CKD sufferers, 99, 102; plantation sickness in the Dooars (India), 24–27, 34–40; soil exhaustion as impact of ginseng farming, 118, 125–26

farmworkers: heat exposure/tolerance among, 105, 108; as migrant labor, 153, 154. *See also* industrial agriculture; *specific countries and crops*

fatigue. *See* exhaustion/fatigue

feminist theory, 7–8, 10–12, 13, 113, 216–17. *See also specific authors*

feral biologies, 26

Ferguson, James, 19n2, 82

figuration, 17, 134, 135–36; and Peruvian gastro-politics, 131, 133, 134–38, 141–42, 143–45

flagellar motors, 171–72, 174–75

folk stories, in Oyama-san's stories of matsutake picking, 192–93, 203–4

food and foraging: caloric intake and energy, 83, 98, 103, 104, 112–13;

food and foraging (*continued*)
  subsistence as human labor regime,
  79, 80, 82–83, 91–92. *See also* agriculture; bees and beekeeping; Indonesian
  orangutan rehabilitation centers;
  sugar
Fortun, Kim, 112
Foster, John Bellamy, 118
Foucault, Michel, 141
Fraser, David, 73–74
Frazer, James, 213
Front for the Defense of Sovereignty and
  Life (Paraguay), 41, 42
Fujimori, Alberto, 143
fungi. *See* matsutake mushroom picking
fungicides: bees and, 161. *See also* chemical use
fusion, as element of Peruvian gastro-
  politics, 132–33, 138, 142, 143

Gaeseong, North Korea, ginseng production and, 116, 121, 127, 129
Galeano, Antonio, 56
García, María Elena, 17, **131–47**
genetically modified crops, 42; Roundup
  Ready crops, 42, 53, 54, 55–56, 163
Ginocchio, Luis, 142
ginseng. *See* Korean ginseng
glyphosate (Roundup) use, 55–56, 57–58,
  163; Roundup Ready crops, 42, 53, 54,
  55–56, 163. See also chemical use
Gordon, Jeffrey, 184, 185–86
Gould, Jeffery L., 103
Green Revolution, in Paraguay, 42,
  44–45, 52, 58
Guardian, on Peruvian food, 144
guinea pig *(cuy)*, and Peruvian gastro-
  politics, 131–47; the cuy in the
  #MásPeruanoQue campaign, 140, *141*;
  the cuy's equation with the nation,
  134, 135, 136, 140, 142, 145–46; the
  cuy's male gendering, 134, 147nn6,13;
  the cuy's refiguration and the current

gastronomic boom, 133, 134–38,
  141–42, 143–45; guinea pig production,
  131, 133, 134–35, 141, 145–46; indigeneity and, 134–35, 136–39, 141, 142, 143,
  144–45, 147n11; the INIA's Guinea
  Pig Project, 131. *See also* Peruvian
  gastro-politics
Guinea Pig Project, 131–32
gut microbes: in honeybees, 161; in
  humans, 185–86. *See also* microbes
  and microbial behaviors

Hackenberg, David, 159
Hall, Stuart, 137
Haraway, Donna, 6, 19n1, 26, 52, 134,
  135, 136
Harris, Marvin, 103
Harrisson, Barbara, 84–87, 93
Harrisson, Tom, 84, 85, 87
Harshey, Rasika M., 171–78, 180–82,
  189n6
Hartigan, John, 18, **171–89**
heat, 97–113; caloric energy, 83, 98, 103,
  104, 112–13; in ethnographic/anthropo-
  logical theories, 103, 112; heat exposure
  and chronic kidney disease, 97–98,
  109–11, 113; and labor, 97–98, 102–5,
  108–9, 112–13; Nicaraguans' experiences of, 98–100, 101–2, 104–5, 107,
  108–11. *See also* Nicaraguan heatscapes/
  workscapes
Helmreich, Stefan, 58n2, 184
Heo Jun, 126
herbicides: glyphosate (Roundup) and
  Roundup Ready crops, 42, 53, 54,
  55–56, 57–58, 163. *See also* chemical use
Hetherington, Kregg, 16, **41–58**, 83
Hobbes, Thomas, 198
Hock, William Gwee Thian, 79
Hodges, Matt, 205
hog farms. *See* American hog farms
Homo faber, 185
honeybees. *See* bees and beekeeping

honey production and markets, 155–57

hongsam (Korean red ginseng). *See* Korean ginseng

human-animal relations: on animals' refusal of work, 217–18, 219–23; at Indonesian orangutan rehabilitation centers, 83–84, 89, 90, 92, 95; on industrial hog farms, 60, 62, 63, 64–69, 73, 74, 75; and Peruvian guinea pig production, 135. *See also* American hog farms; bees and beekeeping; Indonesian orangutan rehabilitation centers; Peruvian gastro-politics

The Human Condition (Arendt), 184–85

Humane Society of the United States, 61

human-fungus relations. *See* matsutake mushroom picking

human health: Indian tea plantation workers, 23, 24, 35, 38, 39; sugar consumption and, 104. *See also* chronic kidney disease

humanism and humanist views, 6, 50–51, 164, 215, 218

human-plant relations: Brazilian migrant farmers and soy in Paraguay, 47–48; Paraguayan campesinos and cotton, 42–46, 48, 51, 58n2; on tea plantations and in other monocultures, 26–27, 32–34, 37–38. *See also* agriculture; industrial agriculture; multispecies relations; *specific countries and crops*

IFC (International Finance Corporation) Compliance Advisor Ombudsman, 107, 110

immigrant labor, in beekeeping, 157

India: the chicken-foot dream/not-dream, 211–13; history of tea cultivation in, 27–31; the slaughterhouse dream/not-dream, 219–21

Indian tea plantations of the Dooars, 16, 23–40; CTC production and consumption, 24, 30–31; history of tea

cultivation in India, 23–24, 27–31; interspecies relations on, 32–34, 37–38; plantation abandonment and sickness, 24–27, 34–40; workers' health and economic statuses, 23, 24, 25, 31–32, 34–38, 39, 49; workers' origins and housing, 29–30, 32, 40

indigeneity and Indigenous peoples, 2; Indigenous orangutan rehabilitation staff in Indonesia, 85–88, 90, 91–93; Indigenous peoples' ejection from the Dooars, 29; and Paraguayan agriculture, 43–44, 53, 54; and Peruvian gastro-politics, 132–33, 134–35, 136–38, 140–45, 146

Indonesian orangutan rehabilitation centers, 1, 17, 79–96; the animals' behavior and experiences, 82, 88–91, 92–93, 95–96; the animals' semi-wildness, 84, 89, 90–91, 95; the *cari makan* or "finding food" metaphor, 79–81, 83, 87–88, 92, 93; early history and staffing issues, 84–88; human-animal relations at, 83–84, 89, 90, 92, 95; human workers' duties, 86, 87, 88–89, 92, 93, 95; working conditions and their impacts on social relations, 81–82, 83–84, 86–88, 92–94, 95–96

industrial agriculture, 153–55; American hog farms, 16, 59–76; economics of, 49–50, 154–55; as organized killing, 52–56, 83; soy farming in Paraguay, 41–42, 46–52, 54–56; tea monocultures of the Dooars, India, 16, 23–40. *See also* agriculture; bees and beekeeping; chemical use; mechanization; monocultures; *specific countries and crops*

inequality, 5, 15, 49, 82, 105, 113

Ingenio San Antonio sugar mill, 100–102, 103, 106–7, 109

INIA (National Institute in Agrarian Innovation), Peru, 131

Inter-American Development Bank, 44

International Finance Corporation (IFC)
Compliance Advisor Ombudsman,
107, 110
interspecies relations. *See* multispecies
relations
intimacy, work-based: on Indian tea
plantations, 26, 27, 32, 37–38, 40n3; on
industrial hog farms, 60, 62, 65–66.
See also human-animal relations;
human-plant relations

Japan: concept of nature in, 196–97;
matsutake mushrooms in, 194–95,
200. *See also* matsutake mushroom
picking

Kawata, Ko, 203, 204
KGC (Korean Ginseng Corporation),
121, 122–23, 124, 125, 129
killing, 41–42; agriculture as organized
killing, 52–56, 83; of bees, in industrial beekeeping, 166; decline of the
Paraguayan *changa* system as, 48–52,
54–55
Kim, Eleana, 17, **115–30**
Kim, Mr. (DMZ EcoResearch staff member), 128–29
Kim Si-Kwan, 121
kinship: metaphorical, of tea workers
and plants, 34; microbes and, 185–86,
187; and substitute labor in orangutan
rehabilitation centers, 93–94, 95
Klages, Ludwig, 201–2
Kohn, Eduardo, 68, 183
Korean ginseng, 17, 115–30; biochemistry
and properties of, 116, 121–22, 124;
cultural associations and celebra-
tions, 123–24, 126–27, 128–30; demand
and markets, 117, 120, 121–24, 126;
farming's environmental/ecological
impacts, 115–17, 118, 125–26, 129; growing and processing of, 116, 118, 120–21,
124–26; as a multispecies assemblage,

119, 120, 129–30; terroirification of, 116,
119–20, 122–29
Korean Ginseng Corporation (KGC),
121, 122–23, 124, 125, 129
Kosek, Jake, 17–18, 57, **149–68**
Kroeber, Alfred, 183

*Labor and Monopoly Capital: The
Degradation of Work in the Twentieth
Century* (Braverman), 50, 62–63
labor value/work ethic, 2–8; in current
political discourse, 3–4; environmental instabilities and, 2–3, 4–5, 9, 10;
feminist and antiracist views on, 11–12,
13, 216–17; paths for questioning, 4–5,
7–8, 9; rejections of, 19n2, 59–60,
216–19; STS theories and, 12–15
labor/work (human): Braverman's views
on, 50, 62–63; as ecological relationship, 6–7, 9–10; finding food as
metaphor for, 79–81; heat and, 97–98,
102–5, 108–9, 112–13; at Indonesian
orangutan rehabilitation centers,
85–88; on industrial hog farms, 59,
60, 61–62, 65–68, 69–73, 74; in Korean
ginseng farming, 124–25, 127; Marx
on, 8–10, 50, 117–18; the nature/labor
or work/labor binary, 6, 8–10, 118–19,
171, 185, 186, 188–89; and the notion
of the multitude, 198; the subject/
object binary, 2–3, 7, 8–10, 166–67,
197; subsistence, 79, 80, 82–83, 91–92;
on usefulness/uselessness, 212–15;
women's labor and feminist theory,
10–12, 13. *See also* labor value/work
ethic; postwork futures
labor/work (nonhuman). *See* multispecies *entries*; nature; nonhuman agency
and productivity; *specific nonhuman
life forms*
La Civilización guaraní (Bertoni), 43
land and landscape: agrarian colonization in Paraguay, 45, 54; capitalist

ruins, 25–26, 27; landscapes as labor regimes, 7; terroirification as value production, 120. *See also* agriculture; environmental instabilities; industrial agriculture; monocultures; nature; Nicaraguan heatscapes/workscapes; terroir

Landecker, Hannah, 122

late industrialism, 112

Latour, Bruno, 137

Lee, Mr. (DMZ EcoResearch staff member), 115, 116, 127–28

Lewis, Marsha, 173–74

Li, Tania Murray, 82

life, 41–42, 48, 58n2; labor as life, 42, 50–51; vitalist views, 43, 44, 50, 58. *See also* killing; Paraguayan agriculture and campesino life

lively capital, 13

Locke, John, 2, 138, 146

Lundu Wildlife Center, 88–89, 90–91, 92–94

Lutalyse, 72–73

Macilvaine, Joe, 152

Malinowski, Bronislaw, 213

Martínez, Virgilio, 137

Marx, Karl, 2, 8–10, 50, 103–4, 117–18, 164, 166–67, 189n3, 217

#MásPeruanoQue campaign (Mistura culinary festival), 139–44

Massumi, Brian, 218, 221

Mateo, Julio, 157

matsutake mushroom picking, 18, 191–209; fungus-human relationship as communication, 197, 199, 201; matsutake biology and taxonomy, 193–94, 195, 209n4; matsutake demand and trade, 195; matsutake habitats and their declines, 194–95, 207; and the notion of the multispecies multitude, 193, 199, 201; pickers' stories and sensory attunement to

nature, 18, 191–93, 199–201, 206–8; temporal coordination/experiences and, 202–8

Matsuyama, Tohey, 176

Matta, Raúl, 142

McWilliams, Carey, 152–53, 154

mechanization, 3, 62, 102, 103–4, 154; in Paraguayan soy agriculture, 49–50, 52, 54–55

Mendes, David, 159

metabolic rift, 10, 17, 118

metabolism *(stoffwechsel)*, 9–10, 83

microbes and microbial behaviors, 18, 171–89; culture medium's influence on swarming behavior, 176–77, 178, 181–82; culturing/culturability of microbes, 173–74, 176; flagellar motor function and output, 171–72, 174–75, 179–80; focusing analysis on boundary work, 188–89; heterogeneity and the division of labor, 177–79, 182, 183; mechanics of swarming, 172–73, 175; microbes as ethnographic subjects, 183–88; microbes in bee bread, 161; quorum sensing behavior, 186–87; the social character of microbial labor, 175, 177–79, 180–82, 184, 185–88; talking about sociality in nonhuman species, 182–83

migrant labor: farmworkers as, 153, 154; honeybees as, 150, 151–52, 153, 155, 156, 158–59, 160, 163–64, 166

migration, 183; as alternative for tea plantation workers, 37–38; Brazilian migrant farmers in Paraguay, 41, 47–48; microbial migrants, 182–83

Miller, John, 160

Mintz, Sidney W., 32–33, 104

Mistura (Peruvian culinary festival), 139–44, *141*

Mittermaier, Amira, 213

monocultures: agricultural work in, 6–7, 27, 29, 31, 54–55, 125; establishment

monocultures (*continued*)
of tea monocultures in India, 28–30;
exhaustion and abandonment of, 6–7,
25–26, 34–38; honeybees and, 152–53,
155, 160, 163; interspecies relations in,
26–27, 32–34, 37–38. *See also* agriculture; industrial agriculture; *specific
countries and crops*
monotony/monotonous labor, 59, 62–63,
74, 91
Monsanto, Roundup and Roundup
Ready crops, 42, 53, 54, 55, 163
Montaigne, Michel de, 215–16
Moore, Jason W., 10, 119
motors, flagellar, 171–73, 174–75
Mukharji, Projit Bihari, 129
multispecies assemblages/ecologies: the
bee's gut as, 161; humans as, 185–86;
Korean ginseng farms as, 119, 120,
129–30; microbial communities as,
178–79, 186–88; the multispecies
multitude, 18, 193, 196–99, 208–9n3;
Paraguayan cotton farming as, 43–44,
58n2; Peru's gastronomic boom as,
134–36; vitalist views, 43, 44, 50, 58n2.
See also multispecies relations; nature;
nonhuman agency and productivity
multispecies ethnography, 14–15, 188
multispecies relations, 9–10; contrasting
temporalities and, 201–3, 206–8; on
Dooars tea plantations and in other
monocultures, 26–27, 32–34, 37–38;
on industrial hog farms, 60, 62, 63,
64–66, 73, 74, 75; matsutake pickers'
sensory attunement to nature, 18,
191–93, 199–201, 206–8; mycorrhizal
mushroom symbioses, 193–94; nonhuman work as communicative practice,
197–99; Oyama-san's stories about
meeting mushrooms, 191–93, 203–5;
the multitude, 197–98; the multispecies multitude, 18, 193, 196–99, 207,
208–9n3

Münzel, Mark, 53
mushrooms. *See* matsutake mushroom
picking

Nading, Alex, 17, **97–113**
National Institute in Agrarian Innovation (INIA), Peru, 131
Native peoples. *See* indigeneity and
Indigenous peoples
nature: an anti-productivist view of, 217;
conventional and contrasting notions
of, 82, 196–97; ecological or metabolic
rift, 10, 17, 118; Marx on labor and, 2,
8–10, 50, 117–18; as multispecies multitude, 18, 193, 196–99, 207, 208–9n3; the
nature/labor or work/labor binary, 6,
8–10, 118–19, 171, 185, 186, 188–89; and
Peruvian gastro-politics, 136–38; in
STS theory, 13. *See also* environmental instabilities; land and landscape;
multispecies assemblages/ecologies;
nonhuman agency and productivity;
terroir
Negri, Antonio, 218
neoliberalism, 3, 46, 218
Nicaraguan heatscapes/workscapes, 17,
97–113; activism of CKD sufferers,
101–2, 104, 106–8, 109–11; chronic
kidney disease and heat, 97–98,
109–11, 113; city dwellers' experiences,
98–100, 102, 108–9; heat and labor,
97–98, 102–5, 108–9, 112–13; Nicaraguans' experiences of heat, 98–100,
101–2, 104–5, 107, 108–11; social relations and, 98, 102, 105, 108–9, 111–13;
sugarcane plantations as, 97, 100–102,
106–8, 109–11
Nicaragua Sugar Estates Limited
(NSEL), 106–8, 109–11
Nieto, Vincent, 171
nonhuman agency and productivity,
5–6, 10, 12–15, 74, 182–83; on animals'
refusal of work, 217–18, 219–23;

Arendt's views on, 118–19, 185, 186;
bees viewed as quintessential laborers,
153–54, 167–68; Marx on, 2, 10, 50;
nonhuman labor as communicative
practice, 197–99; the subject/object
binary, 2–3, 7, 8–10, 166–67, 197;
temporality and, 202–3, 207–8. *See
also* land and landscape; multispecies
entries; nature; *specific nonhuman life
forms*
nonhumans. *See also* multispecies
assemblages/ecologies; multispecies
relations
NSEL (Nicaragua Sugar Estates Limited), 106–8, 109–11

Ogawa, Makoto, 194
operaismo, 217, 218–19
orangutan rehabilitation centers. *See*
Indonesian orangutan rehabilitation
centers
Oyama, Kenji, 191–93, 197, 199–200,
203–6, 208n1

Pacheco, Jorge, 100–102, 104, 106–7, 110
Pacheco, Ulises, 100–102, 104, 107, 110
Paju Gaeseong Ginseng Festival, 126–27
Paraguayan agriculture and campesino
life, 1, 16, 41–58; agriculture as organized killing, 52–56; campesino
activism, 41, 42, 48, 55–56; cotton
development and farming, 42–46,
48, 51, 58, 58n2; crop breeding, 52–53,
58; glyphosate use, 55–56, 57–58; soy
farming and decline of the *changa*,
41–42, 46–52, 54–56; the use of fire, 57
Partridge, Jonathan, 171, 172, 174–80
Paxson, Heather, 120, 126
Pellas Group, 106, 109, 111
Peruvian civil war, 132, 142–43
Peruvian gastro-politics, 17, 131–47;
and the guinea pig's refiguration,
131, 133, 134–38, 141–42, 143–45; and

indigeneity/inclusion, 132–33, 134–35,
136–38, 140–45, 146; the Mistura
festival and the #MásPeruanoQue
campaign, 139–44, *141*; and national
identity and sovereignty, 134, 135, 136,
140, 142–43, 145–46; and Peru's violent
past, 132, 142–43. *See also* guinea pig,
in Peruvian gastro-politics
pesticides. *See* chemical use
pharmakon, 128, 129
Pig Chase, 61
pigs. *See* American hog farms
plantation agriculture, 118. *See also*
Indian tea plantations; monocultures
plants and plant labor. *See* agriculture;
human-plant relations; multispecies
entries; nonhuman agency and productivity; *specific crops*
Plato, 128
Plumwood, Val, 133
political discourse, 3–4, 18–19
post-autonomia, 218
Postone, Moishe, 9
postwork futures and imaginaries, 4,
7–8, 12, 14, 18, 19; compensation
designs to address unemployment,
82; the Situationist Manifesto's view,
214–15. *See also* refusal of work
Povinelli, Elizabeth, 27, 39
productivism and anti-productivism,
9, 217
promissory capital, 13

quorum sensing (microbes), 186–87

Rabinbach, Anson, 102
race: views on multiethnicity and hybrid
vigor, 43–44. *See also* indigeneity
Rappaport, Roy A., 103
red ginseng. *See* Korean ginseng
refusal of work, 18, 211–24; the chicken-
foot dream/not-dream, 211–13; learning from/identifying with animals,

refusal of work (*continued*)
217–18, 219–23; the row-your-boat
metaphor, 211, 223–24; on the tyranny
of ethnographic context, 212–15; on
uselessness/usefulness, 213–16
resistance. *See* activism and resistance
rhythm, tempo, and rhapsody, 201–3,
207. *See also* time and temporality
Richards, Peter D., 53
Rockefeller Foundation, 44
Røder, Henriette L., 188
root rot, in Korean ginseng, 125–26
Rosaldo, Renato, 83, 91
Roundup, 55, 163. See also glyphosate
(Roundup) use
Roundup Ready crops, 42, 53, 54, 55–56,
163
Rousseau, Jean-Jacques, 213, 224
Russell, Edmund, 14

Salazar Parreñas, Juno, 17, **79–96**
Sanabria, Natalia, 52–53
Sarawak Museum, 81, 84, 85, 86, 87, 93
Sarawak orangutan rehabilitation centers. *See* Indonesian orangutan rehabilitation centers
satoyama forest habitats, 194–95, 207. *See
also* matsutake mushroom picking
Satsuka, Shiho, 18, 83, **191–209**
science and technology studies (STS),
12–15, 188
scientific management, 62
Scott, James C., 19n2
Sendero Luminoso, 134, 143, 146n1
sharecropping, 51
Shining Path (Sendero Luminoso), 134,
143, 146n1
shizen ("nature"), 196–97
sideshadows, 205, 208
Simondon, Gilbert, 198
sint'oburi (Korean aphorism), 120, 124
Situationist Manifesto, 214–15
slavery, 51, 53

Sloterdijk, Peter, 213
Smith, Adam, 50
Smithfield Foods, 61
sociality (human): Arendt's view of the
social character of human labor, 185;
gut microbes and, 185–86; heat and,
98, 102, 105, 108–9, 111–13; Indian tea
plantation work and, 31–33, 37, 39, 40;
orangutan rehabilitation work and,
81–82, 86–88, 93–94, 95–96. *See also*
kinship; solidarity
sociality (nonhuman), 183, 185–86; of
microbial behaviors, 172–73, 175,
177–83
Society for Cultural Anthropology Biennial Meeting (2014), 15
solidarity, 178; among Nicaraguan
CKDnt activists, 107–8, 110; heat as
catalyst for, 98, 102–3, 105, 108, 111–13
Soper, Kate, 137
Sørensen, Søren J., 188
South Korean ginseng farming. *See*
Korean ginseng
soy farming in Paraguay, 16, 41–42, 46,
47–58; as organized killing, 52–56,
57–58; as threat to campesino life,
41–42, 48–52
Spinoza, Baruch, 197
Stellabarrie tea plantation, 34–38
stereotypic behavior, 73–74
Stoler, Ann Laura, 141–42
Strathern, Marilyn, 111, 213
Stress and Freedom (Sloterdijk), 213
Stroessner, Alfredo, 44, 45, 46
STS (science and technology studies)
theories, 12–15, 188
subsistence: the *cari makan* metaphor,
79–81, 83, 87–88, 92, 93; as human
labor regime, 79, 80, 82–83, 91–92
sugar, 103, 104, 110; in a Nicaraguan
CKD sufferer's experience, 99, 102,
104; wartime rationing and the honey
industry, 155

sugarcane plantations and workers,
32–33, 102–3, 104; chronic kidney disease in cane workers, 97, 100–102, 104,
106–8, 109–11
Suh, Soyoung, 120
Swettenham, Frank, 79
Syngenta, 48

Tanabata legend, 192–93, 203–4
Tea Cultivation, 28–29, 31, 40n2
tea plantations. *See* Indian tea
plantations
Tegel, Simeon, 144
tempo, rhythm, and rhapsody, 201–3. *See
also* time and temporality
terroir, 120; the terroirification of Korean
ginseng, 116, 119–20, 122–29
Thacker, Eugene, 41, 50
time and temporality: labor and, 81,
202, 204–7; mushroom picking and
temporal coordination/experiences,
202–8; rhythm, tempo, and rhapsody,
201–3
Tracy, Megan, 122
3 de Noviembre, Paraguay, 46, 48. *See
also* Paraguayan agriculture and
campesino life
Tronti, Mario, 217, 218–19
Tsing, Anna, 25–26, 32, 93, 202

Uexküll, Jakob von, 201
unemployment, 19n2, 82
US agriculture. *See* American hog farms;
California agriculture
US unemployment rates, 19n2
unpaid work: domestic work, 11. *See also*
nonhuman agency and productivity

US Agency for International Development, 44
uselessness/usefulness, 212–16, 217, 224.
See also refusal of work

Valderrama, Mariano, 139
Varroa mite, 160, 165
Vasconcelos, José, 43
Virno, Paolo, 198
vitalism, 43, 44, 50, 58n2

Weber, Max, 217
weeds. *See* chemical use; herbicides
Weeks, Kathi, 7–8, 12, 15, 19n2, 59, 75,
216–17, 219
West, Stuart, 186, 187
What Animals Teach Us about Politics
(Massumi), 218
Wilderson, Frank, III, 9
wild mushrooms. *See* matsutake mushroom picking
Wolf, Eric R., 11
wonder, 213–14
work: Arendt's work/labor distinction,
118–19, 185, 186. *See also* labor *entries*;
nonhuman agency and productivity;
refusal of work
Worker in the Cane (Mintz), 32–33
World Bank, 107, 109

Yanagisako, Sylvia, 15, 16, 93
York, Richard, 118

Zaldívar, Marco, 131, 146n1
Zavella, Patricia, 26, 32, 33

www.ingramcontent.com/pod-product-compliance
Lightning Source LLC
Chambersburg PA
CBHW030645270326
41929CB00007B/217